Erwin Schrödinger and the Quantum Revolution

John Gribbin

WILEY

Cover Design: Tom Poland
Cover Image: Professor Erwin Schrödinger © Bettmann/CORBIS

Published by John Wiley & Sons, Inc., Hoboken, New Jersey

First published in Great Britain in 2012 by Bantam Press an imprint of Transworld Publishers

Library of Congress Cataloging-in-Publication Data

Gribbin, John, date.
Erwin Schrödinger and the quantum revolution / John Gribbin.
pages cm
Includes bibliographical references and index.
ISBN 978-1-118-29926-5 (hardback); ISBN 978-1-118-33411-9 (ebk);
ISBN 978-1-118-33188-0 (ebk); ISBN 978-1-118-33519-2 (ebk)
1. Schrödinger, Erwin, 1887-1961. 2. Physicists—Austria—Biography.
3. Physics—Philosophy. 4. Quantum theory. I. Title.
QC16.S265G75 2013
530.092—dc23
[B]

2013025742

Printed in the United States of America

10 9 8 7 6 5 4 3 2 1

For Terry Rudolph,
even though he won't read it

We must not forget that pictures and models finally have no other purpose than to serve as a framework for all the observations that are in principle possible.

ERWIN SCHRÖDINGER, Frankfurt, December 1928

His private life seemed strange to bourgeois people like ourselves. But all this does not matter. He was a most lovable person, independent, amusing, temperamental, kind and generous, and he had a most perfect and efficient brain.

MAX BORN, *My Life* (1978)

Contents

Preface xi
Acknowledgements xii

Introduction: It's Not Rocket Science 1

1 Nineteenth-Century Boy 7
Antecedents—Early years—An empire's last hurrah—Scientific stirrings—From schoolboy to undergraduate

2 Physics before Schrödinger 22
Newton and the world of particles—Maxwell and the world of waves—Boltzmann and the world of statistics

3 Twentieth-Century Man 44
Student life—Life beyond the lab—War service on the Italian Front—Back to Vienna—The aftermath—The peripatetic professor

4 The First Quantum Revolution 67
When black bodies are bright—Enter the quantum—The quantum becomes real—Inside the atom—Tripping the light fantastic—Einstein again

5 Solid Swiss Respectability 91
The university and the ETH—Personal problems and scientific progress—Physics and philosophy—Life and love—"My world view"—Quantum statistics

6 Matrix Mechanics 112
Half-truths—What you see is what you get—Matrices don't commute—Justice isn't always done

7 Schrödinger and the Second Quantum Revolution 124
Science and sensuality—Riding the wave—A quantum of uncertainty—The Copenhagen consensus

8 The Big Time in Berlin 148
Making waves in America—Berlin and Brussels—The golden years—Back to the future—People and politics

9 The Coming of the Quantum Cat 172
Back in the USA—Oxford and beyond—Faster than light?—The cat in the box—From Oxford with love

10 There, and Back Again 187
Whistling in the dark—Reality bites—The unhappy return—Belgian interlude

11 "The Happiest Years of My Life" 201
"Dev"—Settling in—Early days at the DIAS—"Family" life in Dublin—The post-war years—Many worlds

12 What Is Life? 224
Life itself—Quantum chemistry—The green pamphlet—Schrödinger's variation on the theme—The double helix

13 Back to Vienna 243
Farewell to Dublin—Home is the hero—Declining years—The triumph of entropy

14 Schrödinger's Scientific Legacy 259
Hidden reality and a mathematician's mistake—The Bell test and the Aspect experiment—Quantum cryptography and the "no cloning" theorem—Quantum teleportation and classical information—The quantum computer and the multiverse—Quantum physics and reality

Postscript: Quantum Generations 285

Notes 291
Sources and Further Reading 297
Index 305

Preface

While writing my book *In Search of the Multiverse*, I came across a prescient but little-known piece of work by the quantum pioneer Erwin Schrödinger, which pointed the way, had anyone taken notice of it at the time, towards the very modern idea of a plurality of worlds, separated from one another not in space but in some other sense—"parallel universes," in the language of science fiction. There was no suitable way to squeeze this historical dead end into that book, but it reminded me that Schrödinger was a man of many talents, and well worth being the subject of a popular biography—a biography which would give me a chance to dust down that forgotten piece of work and give it the recognition it deserves, set in the context of Schrödinger's life and work. The more I looked into his life, the more remarkable it seemed; I hope you agree that his is very much a story worth telling.

Acknowledgements

Although her name does not appear on the cover of this book, Mary Gribbin played an invaluable role as a researcher, digging up details of Schrödinger's life and liaising with libraries and research institutions. As ever, we are both indebted to the Alfred C. Munger Foundation for financial support. Our special thanks go to the following people who and places that helped us, over the years, in our search for Schrödinger: Michel Bitbol; Dominic Byrne; John Cramer; Dublin Institute for Advanced Studies; Einstein Archive, Princeton; Johns Hopkins University Archive; Sir William McCrea; Oxford University Archive; Rudolf Peierls; Terry Rudolph; Schrödinger Archive, Alpbach; Schrödinger Archive, Vienna; Christine Sutton; University of Berlin Archive; University of Wisconsin Archive; Vienna University Archive.

INTRODUCTION

It's Not Rocket Science

Rocket science is the purest expression of the laws of physics spelled out by Isaac Newton more than three hundred years ago, often referred to as "classical" science. Newton explained that any object stays still or moves in a straight line at constant speed unless it is affected by an outside force, such as gravity. He taught us that if you push something it pushes back—action and reaction are equal and opposite, as when a rifle kicks back against your shoulder while the bullet flies off in the opposite direction. He also gave us a simple law of gravity, explaining how the force of gravity depends on mass and distance. The "action and reaction" bit is at the heart of rocket science. A rocket throws out stuff (usually hot gas, although in principle machine-gun bullets would do the trick) in one direction, and the reaction makes the rocket accelerate in the opposite direction. When the motors are not running, the spaceprobe drifts in what would be a straight line except for the influence of gravity. All sound Newtonian physics, and not really very difficult to understand.

Classical science describes an utterly predictable world. It is possible, for example, to work out exactly how much rocket thrust in what direction is needed to set a spaceprobe with a certain mass on a trajectory, falling through space under the influence of gravity, that will take it to intercept the planet Mars at a precise date months in the future. Assuming their engines work properly, spaceprobes only miss their target when somebody gets the sums wrong—when there is human error.

For centuries after the time of Newton classical science posed a real problem for anyone who believes in free will. In principle, if you knew the position and speed of every particle in the Universe, including the atoms we are made of, at any chosen moment of time, it would be possible not only to predict the entire future of the Universe, but to reconstruct its entire history in exquisite detail. Leaving aside the practical problems of actually doing this, it seemed to imply that everything, including human behaviour, was pre-ordained. But then came quantum physics.

Quantum physics is not like classical physics. It is definitely not rocket science; it's much harder to understand than that. It took many top scientists, working over the first three decades of the twentieth century, to work out just what quantum physics is, and when they did find out some of them, including the subject of this book, didn't like what they had found.

Quantum physics mostly describes the world of the very small—roughly speaking, things the size of atoms and smaller. What physicists painstakingly (and painfully) discovered during those first three decades of the twentieth century is that particles can behave like waves and waves like particles; that quantum entities can be in at least two places at once; that they can get from one place to another without passing through any of the space in between; and that there is no certainty in the quantum world, where everything depends on probabilities. It's as if you sent a spaceprobe on its way in the knowledge that there was a 50 per cent chance that it would arrive at Mars and a 50 per cent chance that it would arrive at Venus, but no way to tell in advance where it would end up. Great for restoring a belief in free will, but scarcely reassuring in any other way. And yet all this baffling behaviour of the quantum world has been tested and confirmed in countless experiments.

Erwin Schrödinger's masterpiece, the work for which he received the Nobel Prize, tried to restore the common sense of classical physics to the quantum world. It isn't giving away too much of our story to say that he failed, and that his work became an integral part of the revolutionary new physics.

But there was much more to Schrödinger than the reluctant revolutionary of quantum physics. One of the most intriguing aspects of Schrödinger the physicist, and one that lies at the heart of his antipathy to the revolution he participated in, is that although he made a major contribution to the new science of the twentieth century, he was brought up in the scientific tradition of the nineteenth. He graduated from high school and started at university in 1906, the year after Albert Einstein published his classic papers on the special theory of relativity and quantum physics. But Einstein, of course, was an exception; his ideas on quantum physics, in particular, were not taken seriously for at least another ten years, and the real quantum revolution took place at the hands of Young Turks such as Werner Heisenberg (born in 1901) and Paul Dirac (born in 1902), who, along with the likes of Niels Bohr, Louis de Broglie, and Einstein, all come into the story of Schrödinger's life and work.

Schrödinger wasn't just a physicist. He was a disciple of Arthur Schopenhauer, with a profound interest in philosophy and Eastern religion, particularly espousing the Hindu Vedanta and subscribing to the idea of a single cosmic consciousness of which we are all part. He studied colour vision, and wrote a book, *What Is Life?*, which Francis Crick and James Watson each independently cited as a major influence on the work which led them to the discovery of the DNA double helix. Schrödinger also addressed questions such as "What is a law of nature?" and whether or not the world is in principle completely deterministic and predictable. He wrote poetry (badly) and a book about the science and philosophy of ancient Greece.

Schrödinger's private life was no less interesting. Brought up in comfort in the last days of the Austro-Hungarian Empire, he served as an artillery officer in the First World War and suffered the consequences of the post-war blockade of Austria (a long-forgotten Allied atrocity which caused mass starvation) and the runaway inflation of the early 1920s. After these experiences, one of his main concerns was to secure his own and his wife's financial future; he worried about pensions until his death. His first attempt to get away from Nazi-influenced Europe came to nothing when he turned up in Oxford with both his wife and his mistress, offending the academic establishment there by making no attempt to conceal their living arrangements, with which his wife, who had her own lovers, was quite happy. The possibility of a post in Princeton alongside Albert Einstein fell through for the same reason. Schrödinger eventually landed up in more tolerant Dublin, where at the behest of Éamon de Valera, the Taoiseach (Irish Prime Minister), the Dublin Institute for Advanced Studies was set up to provide him with a base.

Schrödinger was also unconventional in other respects. As a university lecturer in the last days of Prussian formality, he neglected to wear a tie, and dressed so casually that he was often mistaken for a student and sometimes for a tramp. On at least one occasion he had difficulty gaining access to an important scientific meeting because he had hiked to the venue, rather than going by train, and presented himself straight off the road, dressed for rambling and carrying a rucksack.

When he retired in 1956, Schrödinger returned to Vienna and served as Austria's representative at the International Atomic Energy Agency before his death in 1961. Like other elderly physicists, including Einstein, he had tried unsuccessfully to find a unified theory of physics. But generations of physics students know him from the equation which bears his name, and countless non-physicists know him from the parable of Schrödinger's cat. The whole point of that parable was to demonstrate the absurdity of quantum physics, and it could only have been dreamed up by a physicist steeped in the classical tradition. So the search for Schrödinger begins with classical physics.

Chapter One

Nineteenth-Century Boy

Erwin Schrödinger was the only child of a wealthy Viennese family in the last decades of the Austro-Hungarian Empire. This upbringing naturally affected the kind of person he grew up to be; it also affected the way he thought about science and influenced the development of his greatest scientific idea, for which he received the Nobel Prize.

Antecedents

Erwin was the son of Rudolf and Georgine (Georgie) Schrödinger, who married in 1886. Rudolf's parents were profoundly affected by the almost casual way in which death could strike even in the most affluent parts of the civilized world in the nineteenth century. At the time of her marriage in 1853 his mother, Maria, was a nineteen-year-old orphan. Just five years later, she died following the birth of a stillborn baby. She had already produced a son, Erwin, who died as a child, a daughter, Marie, and another son, Rudolf, born on

27 January 1857. Her husband, Josef, whose family came originally from Bavaria but had lived in Vienna for several generations, brought up the surviving children on his own, without (as would have been more usual at the time) remarrying. But although the children may have lacked a mother, their material needs were well catered for. Josef owned a modest but profitable business, a factory manufacturing linoleum and oilcloth; this family business would in due course be passed on to his surviving son, Erwin's father, Rudolf.

Socially, Georgie's family were a cut above the Schrödingers; indeed, they had aristocratic pretensions. They were descended from a minor nobleman, Anton Wittmann-Denglass, who had been born into a Catholic family in 1771. Such were the religious strictures of the time that when his daughter Josepha fell in love with a Protestant, she was forced to abandon her love-match and marry the family doctor, a good Catholic. She had three children before, perhaps to her relief, she was widowed and able to marry again. This time she chose—or had chosen for her—Alexander Bauer, the manager of her father's estates. The eldest son of this second marriage, another Alexander Bauer, was born in 1836. He would become Erwin Schrödinger's maternal grandfather. Alexander Bauer was the first in the family to show an interest in science, studying mathematics and chemistry in Vienna and Paris, and moving on to become a research chemist.

Erwin's maternal grandmother, Emily, was English and also came from a family with upper-class connections. They claimed descent from the Norman Forestière family, although the name had long since been anglicized to Forster, and had been associated with Bamburgh Castle in north-east England.

Thomas Forster, born in 1772, was the son of the governor of Portsmouth, and his eldest daughter, Ann, born in 1816 and one of five children, would become Erwin's great-grandmother. He had met her when visiting England as a child. Ann married a solicitor, William Russell, and they had three children—William, Emily, and Ann (known as Fanny).

The younger William Russell became an analytical chemist. In 1859–60, while studying chemistry in Paris, he met fellow student Alexander Bauer. The two became friends, and when Emily (nicknamed Minnie) and her mother visited William in France, Alex met Minnie, then just nineteen years old, and the couple fell in love. Once Alexander had completed his studies and obtained his first (very junior) academic post, they were able to marry. After their wedding in Leamington Spa, on 21 December 1862, they lived in Vienna, where their first daughter, Rhoda, was born in 1864, followed by Georgie in 1867; soon after the birth of a third daughter, another Emily/Minnie, in 1874, Emily died of pneumonia.

Alexander Bauer's career continued to flourish until 1866, when he lost an eye in an explosion at the laboratory. From then on he concentrated on teaching, his studies in the history of chemistry, and the inevitable administrative duties associated with his rise to become Professor of General Chemistry at the Vienna Polytechnic (later the Technical University of Vienna), a post he held until his retirement in 1904. He was also a curator of the Museum of Art and Industry and a member of the Theatre Commission for Lower Austria, and took pleasure in introducing his grandson Erwin to the theatrical arts at an early age.

Alexander was devoted to his daughters, all of whom married men they had met through their father's connections.

Rhoda, the eldest, married the Director of the Viennese Pharmaceutical Commission, Hans Arzberger, but had no children. Minnie, the youngest, married Max Bamberger, who later succeeded Alexander as Professor of General Chemistry, and had a daughter, Helga. Georgie married Rudolf Schrödinger.

Rudolf was a frustrated scientist who had studied under Alexander Bauer at the Technical University, but was obliged to take over the family business rather than pursue a career in chemistry. He married Georgie on 16 August 1886, when he was twenty-nine and she was nineteen. Although Rudolf, like most Austrians, was at least nominally a Catholic, the wedding took place at a Lutheran church (Georgie and her sisters had been brought up in the Lutheran tradition, the nearest thing in Austria to the Anglican religion of their mother), making their son Erwin nominally a Protestant, although as we shall see this meant little in practice. The family was essentially irreligious, attending church only for weddings and funerals. Indeed, when Erwin Rudolf Josef Alexander Schrödinger, named after his father's dead brother, his father, and his two grandfathers, was born in Vienna on 12 August 1887 and baptized five days later, even the naming ceremony took place at the Schrödingers' home, not in church.

Early years

Although Erwin's English grandmother had died thirteen years before he was born, her influence on the Schrödinger family was strong. His aunt Rhoda had grown up hearing only English spoken at home, and had spent years with her own grandparents in Leamington Spa. His mother's younger sister,

Minnie, who was similarly fluent in English, was only fourteen years older than Erwin, and played with him as a child. So Erwin grew up hearing both English and German spoken at home; according to some accounts, he spoke good English before he learned to speak "proper" German.

Erwin was an only child with two doting aunts, a female first cousin (Dora, the daughter of his father's sister), and a succession of nurses and maids attending almost to his every whim. It is tempting to see here the origin of patterns in Erwin's adult relationships with women. He grew up to expect women to dance attendance on him, while being somewhat insensitive to their needs. According to the psychiatrist Dennis Friedman, a boy brought up with both his mother and a nanny to look after him is predisposed to become a philanderer in later life: the experience

> creates a division in his mind between the woman he knows to be his natural mother and the woman with whom he has a real hands-on relationship: the woman who bathes him and takes him to the park and with whom he feels completely at one . . . he grows up with the idea that although he will one day go through all the social and sexual formalities of marriage, he will have at the back of his mind the notion of this other woman, who not only knows, but caters for, all his needs.[1]

Although this suggestion has been challenged (for example, by child psychologist Linda Blair), Friedman could have used Schrödinger as a case study in support of his hypothesis. But any such consequences lay far in the future when the boy Erwin was growing up in Vienna.

At the time Erwin was born, his grandfather Alexander owned a new town house in the centre of Vienna. The five-storey building was divided into five separate apartments, and in 1890 "our" Schrödinger and his parents moved in to the spacious fifth-floor accommodation, with views overlooking St. Stephan's Cathedral.

Most of what we know about Erwin's early life comes from the recollections of his aunt Minnie, which should be taken with the same pinch of proverbial salt as similar recollections made (much later in life) by relatives of Albert Einstein about his precocious childhood. But in both cases the reminiscences surely contain seeds of truth. From an early age, Erwin was interested in astronomy: he would persuade Minnie to stand representing the Earth while he ran round her to be the Moon, and then make her walk in a circle around a light representing the Sun while he continued to run round her. He also kept a kind of daily diary even before he could write, dictating his insights to Minnie. A surviving entry from 1891 reads: "In the evening Aunt Emmy [Minnie] cooked a good supper and then we spoke all about the world." Recording his thoughts and activities on paper was to become a lifetime habit.[2]

Erwin did not have to leave his cosy family circle even to go to school until he was ten, since up to that time he was tutored privately at home for two mornings a week. According to Minnie, he began to read almost as soon as he could talk, thanks to a maid who explained the names on street signs to him; but apart from such basics, the purpose of his early tuition was to prepare him for the entrance examination for the *Gymnasium* (equivalent to an English grammar school), where his real education would begin. But while the

Schrödingers enjoyed the stereotypical life of the upper middle classes in Vienna, the empire around them was showing signs of the strains that would soon alter all of their lives, not least the young Erwin's, for the worse.

An empire's last hurrah

Vienna had been the capital of a great empire for centuries, ruled since 1276 by the Habsburg family. The geographical extent of this empire varied considerably over time. During the sixteenth and seventeenth centuries its fortunes ebbed and flowed, and in 1683 the expanding Ottoman Empire reached as far as Vienna before being repulsed. But even after the incursions of the Napoleonic Wars the Emperor (then Franz) ruled not only over much of the German-speaking world but also over Hungary, much of Poland, and what became Czechoslovakia, parts of Italy, and, crucially for European history, the Slavic states in Balkan Dalmatia.

Towards the end of the eighteenth century the French Revolution had started a fire that would slowly spread across Europe to end the age of the great European empires. In 1848 the Continent was rocked by a series of political upheavals so widespread and significant that it became known as "the year of revolutions." In the Austrian Empire, risings in Italy, Bohemia, Hungary, and Vienna itself were put down by force, but concessions had to be made; the Emperor, Ferdinand (who had succeeded Franz in 1835), was forced to abdicate.

The new Emperor was Ferdinand's nephew, Franz Josef, who had been born in 1830 but, in spite of his youth, at first looked backward rather than forward, dreaming of re-creating an absolute monarchy ruling over a strong, expanding Austrian Empire. The harsh reality of military and political failures, including the Crimean War and the loss of Lombardy and Venice, forced him to change his approach, and from the mid-1860s onward Franz Josef became less autocratic and granted his people a greater degree of freedom. In 1867 Hungary achieved (at least nominally) equal status with Austria in what became called the Austro-Hungarian Dual Monarchy or more succinctly Austria-Hungary. But while the empire lost some territory, it gained elsewhere. In 1878 it took over the administration of the Balkan states Bosnia and Herzegovina, although these remained nominally part of the Turkish Empire until Austria-Hungary annexed them in 1908.

So the Vienna in which Erwin Schrödinger was raised was the capital of an empire that was visibly fraying at the edges. It contained people of many different nationalities and political allegiances, many of them dreaming of, or working for, independence. This was also, of course, a time of great social change, with industrialization, improved communications, and the consequent movement of people into the cities. Increasingly, as Franz Josef aged he became a relic of times long gone, losing his power and influence to a bureaucratic system which rumbled on as much through inertia as anything else.

Vienna, to some extent insulated from these realities, remained a glamorous city famous for the arts. The Viennese loved opera and music, and in the nineteenth century the tradition of Haydn, Mozart, and Beethoven was carried forward here by Schubert, Liszt, Brahms, and Bruckner. And, of course, the Strauss family. But the people who now enjoyed these cultural delights were increasingly, like Rudolf Schrödinger, the new bourgeoisie, rather than the old aristocratic class. Among the most important of these upwardly mobile groups were the Jews. Like all non-Catholics in Austria, they had had few rights (let alone privileges) before 1848, but as the grip of the authorities eased, Jews from all over the empire were among the people attracted to the capital. They made an economic and artistic impact out of all proportion to their numbers, in a society where casual anti-Semitism was common and "the Jews" often got the blame for anything wrong with society. But this was not a prejudice that Erwin would grow up to share.

Scientific stirrings

Already famous for the arts, Austria, and Vienna in particular, also had a burgeoning scientific reputation in the second half of the nineteenth century. Among the changes that took place after 1848 was the establishment of a Physics Institute at the University of Vienna; its Director, Johann Christian Doppler, also became the first Professor of Experimental Physics at the university. Born in Salzburg and educated in Vienna, Doppler served in academic posts around the Austrian Empire before being appointed to head the new institute. Although he also did important work in mathematics and the study of electricity, he is remembered today for his investigation of the way in which the pitch of a sound or the colour of a light is affected by the relative movement of the source and the person observing it. His calculations were famously confirmed in 1845 by the Dutch meteorologist Christoph Buys Ballot, who arranged for trumpeters standing on an open car blowing a single note with all their might to be towed by a train past musicians with perfect pitch standing at the side of the track listening to the change in the note they heard as the train, and the trumpeters, passed them. It is this "Doppler effect" that explains the change in pitch of the siren of an emergency vehicle as it rushes past, and its optical equivalent is used to measure the speed with which stars are moving towards or away from us.

Doppler died in 1853 at the age of only forty-nine, and was succeeded as Director of the Physics Institute by Andreas von Ettinghausen. When von Ettinghausen, an undistinguished scientist, became ill in 1862 an acting director had to be appointed; and the man chosen was the rising star of Viennese physics, twenty-seven-year-old Josef Stefan, then a junior member of the university (a *Privatdozent*, the first rung on the academic ladder). Stefan became a full professor a year later, and in 1866 was officially appointed Director of the Institute. A pioneer in the study of thermodynamics (discussed in Chapter 2), Stefan investigated the way electromagnetic energy (heat and light) is radiated from a hot object. His findings, refined by his student Ludwig Boltzmann (himself Viennese), became known as the Stefan–Boltzmann Law of black body radiation, a key step on the road to what became the first version of quantum physics.

As well as being a first-rate scientist, Stefan was also a first-rate teacher, and one of his students, Fritz Hasenöhrl, would have a profound influence on Erwin Schrödinger, so that in an academic sense Stefan was Schrödinger's "grandfather." Schrödinger's other academic grandfather was Stefan's colleague Josef Loschmidt, whose particular claim to fame lay in calculating how molecules bouncing off the walls of a container produce pressure, thereby convincing his contemporaries of the reality of molecules, although he did a great deal more work in the young science of thermodynamics. His student

Franz Exner, who was to succeed him as professor at the university (and was, incidentally, instrumental in persuading the Viennese authorities to give Marie and Pierre Curie the pitchblende in which they discovered radium), was Schrödinger's mentor in experimental physics, while Hasenöhrl was his mentor in theoretical matters.

By the end of the nineteenth century, physics was thriving in Vienna. But while the Austrian physicists were honoured in their own country, their renown was much less abroad—not least because Stefan and Loschmidt, in particular, never travelled to spread the word about their achievements. As Boltzmann commented in 1905:

> Neither Stefan nor Loschmidt went, according to my knowledge, on a travel beyond the borders of [their] Austrian homeland. At any rate, they never visited a [scientific conference] and did not establish closer personal relationships with foreign scientists. I cannot approve of that, for I believe that they could have achieved still more if they had closed themselves off less. At least they would have made their achievements known faster.[3]

This was not a mistake that Boltzmann himself made. He led the way in promoting Austrian—or at least, his own—achievements among the wider scientific community. Boltzmann appreciated that by the end of the nineteenth century science was an international endeavour in which it was essential to maintain contact with colleagues in different countries. Nobody would epitomize the international nature of physics in the twentieth century better than Erwin Schrödinger, who arrived at the university just a year after Boltzmann made those remarks.

From schoolboy to undergraduate

He might have entered university in 1905, but Erwin's formal education was delayed by a year because he sat the entrance examination for the *Gymnasium* later than usual, having taken a long holiday in England with Georgie and her sister Minnie in the spring of 1898, when he was ten. It was on this trip that he met his great-grandmother Ann, who had been born the year after Napoleon's final defeat at the Battle of Waterloo. Minnie tells us that he also learned to ride a bicycle, rode a donkey on the sandy beach at Ramsgate, and visited a great-aunt who kept six Angora cats.

The holiday did not end when the party left England from Dover by steamer. From Ostende they travelled to Bruges and Cologne, and then up the Rhine by boat as far as Frankfurt-am-Main before completing the journey home by train. In order to prepare him for the examination (which he passed with ease), Erwin briefly attended St. Niklaus School, his first experience of formal education. He entered the *Gymnasium* in the autumn of 1898, a few weeks after his eleventh birthday. The school was the most secular one of its kind in Vienna, and numbered Boltzmann among its former pupils. But probably neither factor played a part in the Schrödingers' choice; more relevant was the fact that it was just ten minutes' walk from the family home, on the Beethoven Platz. Erwin would be a pupil there for the next eight years.

The *Gymnasium* offered a classical education dominated by the study of Latin and Greek language and civilization. Lessons took place from 8 a.m. until 1 p.m., six days a week, with an additional two afternoons a week for the study of Lutheranism. "From this," says Schrödinger, "I learned many things, but not religion." His favourite question was, "Sir, do you really believe that?"

In the first three years of school, there were eight hours of Latin a week, reduced to five hours when the pupils started Greek. There were also courses in German language and literature, geography, music, and history. All of this left just three hours a week for mathematics and science. Hardly surprisingly, the maths never got as far as calculus, but covered the mathematical equivalent of the classics: geometry and algebra. Much of the physics would have been familiar to Newton, and although there were classes in biology (mostly botany), the only mention made of Darwin's theory of evolution by natural selection was in the religious class, where it was denounced. The young Schrödinger learned more about the natural world from his father, a keen amateur botanist who published articles in learned journals (and who always regretted having had to give up an academic career for the family business); but he, too, was cautious about accepting all of Darwin's ideas. One of Rudolf's friends, however, was a zoologist at the Natural History Museum, and much more enthusiastic about natural selection. Under his influence, Erwin soon also became an enthusiastic Darwinian.

It was while he was at the *Gymnasium* that the first signs of Erwin's remarkable intelligence became evident outside his family. He was a good student who tells us that he loved mathematics and physics, but also enjoyed the logic of grammar and philology; he loved poetry but hated "the pedantic dissection" of literature. He was always top of the class, in every subject, and a schoolmate later recalled the deep impression made on the pupils by the only occasion on which Erwin failed to answer a question from the teacher—it was "What is the capital of Montenegro?"[4] In the afternoons (when not required to attend religious instruction), Erwin studied English and French. Perhaps because of his upbringing in a bilingual household, he became an excellent linguist who could lecture in German, French, English, and Spanish, switching between them to answer questions from polyglot audiences; as an adult he also translated Homer into English and Old Provençal poetry into modern German.

The student who had the frustrating experience of being second in class to Erwin throughout his eight years at the *Gymnasium* was Tonio Rella. In spite (or perhaps because) of this, they became firm friends. The Rella family owned an inn in the mountainous countryside, and Erwin often spent holidays there, developing, with Tonio, his love of hiking and the outdoor life. He also developed his first adolescent crush, on Tonio's sister Lotte, although in the circumstances of the time this could not develop much beyond holding hands. Tonio went on to become Professor of Mathematics at the Viennese Technical University, and remained friends with Erwin; he was killed by the shelling during the Russian advance on Vienna at the end of the Second World War.

Erwin's other great love as an adolescent was the theatre, one of the highlights of Viennese life at the beginning of the twentieth century. He usually went at least once a week, often taking advantage of the special matinees held on Sunday afternoons for students and workers. The main theatre was the splendid Hofburg, on the Ringstrasse, one of the most important German-language theatres in the world; but even the lesser Volkstheater could seat 1,900 people, while smaller theatres offered operetta, farce, and even Hungarian vaudeville. Always an obsessive note-taker, Erwin even kept a record, with mini-reviews, of his visits to the theatre. Of one leading actor, he wrote: "Much better than I had expected, not so much by what he does as by what he has left undone."

The visual arts were also at a peak in turn-of-the-century Vienna, although not always appreciated. Gustav Klimt, at the height of his powers, was creating a storm of controversy with what were perceived as his overly sexual, even pornographic, paintings. In 1906, the year that Erwin Schrödinger and his friend Tonio Rella entered university, Egon Schiele, one of Klimt's friends, was gaoled for twenty-four days for painting a "lewd" picture. But if Vienna was at the cutting edge of theatre and art in 1906, it was certainly not at the cutting edge of physics. While Stefan and Boltzmann were beginning to make interesting advances in their own research, the teaching of physics was lagging sorely behind. Most of what Schrödinger would learn at the university as an undergraduate had a distinctly old-fashioned flavour, even in 1906.

Chapter Two

Physics before Schrödinger

The physics that Schrödinger learned as an undergraduate rested, like a tripod, on three legs: the understanding of mechanics developed by Isaac Newton; the understanding of electromagnetism developed by James Clerk Maxwell; and the understanding of thermodynamics to which Ludwig Boltzmann was a major contributor. He was taught nothing about the new ideas of Albert Einstein, whose special theory of relativity was only published in 1905, and very little about Max Planck's investigation of electromagnetic radiation, published in 1900, which came to be seen as the birth of quantum theory. For our purposes, physics before Schrödinger means physics before 1900. And it begins with Isaac Newton.

Newton and the world of particles

Isaac Newton (1642–1727) is widely regarded as the founder of modern science. This is true in the sense that he spelled out the mathematical laws which describe the motion of objects, and realized that the same laws which govern the behaviour of objects here on Earth—in particular, the law of gravity—govern the behaviour of the Universe at large. This realization of the universality of the laws of physics was far more important than the discovery of the laws themselves. It meant that scientists could expect, eventually, to explain everything in the Universe on the basis of laws that could be investigated in their own laboratories.

But even Newton did not do everything on his own. Right at the beginning of the seventeenth century, the English physician and scientist[1] William Gilbert (1544–1603) published a treatise on magnetism, *De magnete*, in which he not only gave a description of magnetic phenomena that was unsurpassed for two hundred years, but extended the understanding derived from his laboratory studies to explain the Earth's magnetic field—a significant step out into the cosmos at that time. Gilbert also spelled out the basis of what became the scientific method: testing hypotheses by experiment and observation, and rejecting any ideas which do not stand up to those tests. Bizarre though it may seem to us, even in Gilbert's day it was still common for philosophers to try to settle arguments about what we would regard as scientific matters—such as whether a heavy object falls faster than a lighter object—literally by argument, rather than by doing experiments. Gilbert was scathing about such people:

> Every day, in our experiments, novel, unheard-of properties came to light . . .
>
> But why should I, in so vast an ocean of books whereby the minds of the studious are bemuddled and vexed—of

> books of the more stupid sort whereby the common herd and fellows without a spark of talent are made intoxicated, crazy, puffed up; and are led to write numerous books and to profess themselves philosophers, physicians, mathematicians, and astrologers, the while ignoring and contemning men of learning—why, I say, should I add aught further to this confused world of writings, or why should I submit this noble and (as comprising many things before unheard of) this new and inadmissible philosophy to the judgment of men who have taken oath to follow the opinions of others, to the most senseless corrupters of the arts, to lettered clowns, grammatists, sophists, spouters, and the wrong-headed rabble, to be denounced, torn to tatters and heaped with contumely. To you alone, true philosophers, ingenuous minds, who not only in books but in things themselves look for knowledge, have I dedicated these foundations of magnetic science—a new style of philosophizing.

And he summed up the "new style of philosophizing" thus:

> In the discovery of hidden things and in the investigation of hidden causes, stronger reasons are obtained from sure experiments and demonstrated arguments than from probable conjectures and the opinions of philosophical speculators of the common sort.

That notion of "sure experiments and demonstrated arguments" is the basis of science.

Of course, the person who is usually credited with developing the scientific method, and in particular with doing experiments with falling bodies, is Galileo Galilei (1564–1642)—although as it happens Galileo himself did not drop objects from the Leaning Tower of Pisa. He did experiments with balls rolling down inclined planes, and also interpreted the famous Leaning Tower experiment, actually carried out by a rival trying to disprove Galileo's claim that a light object and a heavy object dropped at the same time would hit the ground together. Where, though, did Galileo learn the scientific method? He was certainly capable of working it out for himself; but if he needed any prodding in the right direction, he definitely got it. We know that he read *De magnete* from approving comments he made about Gilbert's book in a letter; and, of course, Newton was thoroughly familiar with the work of Galileo and others such as René Descartes (1596–1650). As Newton himself put it, "If I have seen farther, it is by standing on the shoulders of Giants." But he certainly did see farther.

Newton studied at the University of Cambridge, and became a Fellow of Trinity College in 1667. Just two years later he was appointed as only the second Lucasian Professor of Mathematics. This gave him security for life, and in those days there was no obligation or even pressure to publish scientific discoveries. Newton mostly preferred to keep his ideas to himself, rather than be bothered with the attention and time-consuming correspondence that would result if they became widely known. One idea he did announce, however, was his invention of a new kind of telescope, which resulted in his being elected a Fellow of the Royal Society (founded in 1660, with the epithet "Royal" bestowed the following year) in 1672. This led to Newton presenting his ideas about light and colours to the Society, and in turn to a virulent argument with Robert Hooke (1635–1703), the man who as Curator of Experiments and later Secretary did more than anyone to make the Society a success. The experience confirmed

Newton's view that publicizing his ideas only led to trouble, and he retreated into his shell in Cambridge. There he continued thinking deeply about the nature of the physical world, but stopped telling anyone about his thoughts.

That changed in 1684, when Edmond Halley (1656–1742) visited Newton in Cambridge. The purpose of his visit was to ask if Newton could help with a problem that had been puzzling Halley, Hooke, and another Fellow of the Royal Society, Christopher Wren (1632–1723). The three scientists had realized that the orbits of the planets around the Sun could be explained by a force which falls off in proportion to the square of the distance of a planet from the Sun (an inverse-square law), but they could not prove that all of the laws of planetary motion, described by Johannes Kepler (1571–1630), must result from such a law. Newton, never one for false modesty, told Halley that he had solved that puzzle long ago, but claimed he could not find the relevant document among his papers, and promised to send a copy to Halley later. It is clear from Newton's surviving papers that this claim was a lie intended to buy time while Newton, confident in his own abilities, actually did solve the problem.

In fact, in 1684 most of Newton's ideas were incomplete, and very little had been fully worked out. Halley's visit was the catalyst which encouraged him to pull everything together into a coherent whole. In order to explain how gravity affected planets, Newton had to produce a mathematical explanation of how forces affect the motion of objects in general, including an understanding of the nature of mass itself, and the way in which an object resists being pushed around—its inertia. In November 1684, Halley received from Newton a nine-page document with the title *De motu corporum in gyrum* ("On the motion of bodies in an orbit"); but this was scarcely more than a throat-clearing exercise, because Newton had become gripped by the idea of preparing a complete description of the workings of the physical world. For about eighteen months beginning in August 1684 he worked obsessively on the project, which became his famous book *Philosophiae naturalis principia mathematica*, usually known simply as the *Principia*, published in 1687.

Newton's description of the physical world rested on three laws of motion, but equally importantly on his concept of inertia and its relation to the mass of an object. His first law simply states that an object stays at rest or keeps moving in a straight line unless it is acted upon by an outside force. It sounds simple—but contained within this law is the significant scientific idea of an "ideal" object moving in some ideal force-free space. On Earth, nothing keeps moving unless it is pushed. Objects fall to the ground, or stop moving along the ground. Newton, like Galileo before him, realized that this is because objects are experiencing external forces such as friction or gravity. In an ideal situation, an object moving freely through space without any external forces would indeed keep going for ever in a straight line. But how would such an object know if it were moving or not? Newton himself thought that there must be some fundamental or "absolute" space against which all motion could be measured, and some fundamental absolute time ticking away the history of the Universe; but, as we shall see, those ideas have since been challenged.

Newton's second law explains how a force alters the motion of an object. The acceleration produced by the force is equal to the force divided by the mass of the object. The change in motion is proportional to the force, and the resistance to change (inertia) is measured in terms of the mass of the object. Before Newton, there was no clear idea of what mass is. It was Newton who defined the concept. In his own words: "The quantity of matter is that which arises conjointly from its density and magnitude. A body twice as dense in double the space is quadruple in quantity. This quantity I designate by the name of body or of mass."

Since Newton's law of gravity tells us that the force exerted on an object is proportional to its mass, while his second law of motion tells us that the acceleration of an object is proportional to the force divided by the mass, together they explain why all objects fall at the same rate, regardless of their mass—the mass cancels out of the calculation. If an object is twice as massive, it needs twice as much force to produce the same acceleration—but it feels twice as much force!

Newton's third law tells us that when one object exerts a force on another, it experiences an equal and opposite force in return. When the Sun pulls on the Earth, the Earth also pulls on the Sun; when an apple is pulled by the gravity of the Earth, the Earth feels an equal force pulling it towards the apple; and so on. The force of gravity acting between two objects, Newton explained, is proportional to their two masses multiplied together and divided by the square of the distance between them.

Another key feature of Newton's work is that he recognized that gravity is a universal force—that every object in the Universe attracts every other object in the Universe in accordance with the same inverse-square law. This was the beginning of the realization that the laws of physics derived from studies here on Earth could be applied anywhere in the Universe: a generalization on a scale undreamed of by previous philosophers. Newton was claiming not only that the laws of physics were universal, but that they could account for differences, even minor differences, between the behaviour of real objects and the behaviour of ideal objects—for example, the way in which friction prevents an object moving in a straight line at a constant speed for ever. It was the beginning of truly quantitative science, and it led to a profound, if puzzling, implication.

When two moving objects collide with one another, Newton's laws enable you to work out exactly how they will bounce apart—in which direction each object will move, and at what speed. The key to this calculation is knowing the speed, direction of motion, and mass of each object at the moment of impact. These three entities are bound up in a single property, the momentum of the object. Speed just tells you how fast an object is moving, not its direction; the velocity is the speed an object has in a certain direction. So if an aircraft is travelling at 500 km/hour, that is its speed; but if I say it is flying due north at 500 km/hour, that is its velocity. The momentum of an object is its mass multiplied by its velocity.

It is worth repeating at this juncture a point made earlier: that a combination of Newton's laws of motion and the inverse-square law of gravity means that in principle, if you knew the position and the momentum of every object in the Universe—every particle, in scientific language—at a certain moment of absolute time, you could not only predict the entire future of the Universe, but also reconstruct its entire history.[2] It doesn't matter that it is impossible for us to do this in practice, since the Universe itself "knows" where everything is and where it is going. The implication is that free will is an illusion, and that everything is predetermined. This leads to the idea of the Universe as a kind of cosmic clockwork train, wound up by God in the beginning and set ticking along its already laid railway track for eternity. Although seldom openly acknowledged, this disturbing implication remained embedded in physics until the quantum revolution of the 1920s.

Newton did very little scientific work after 1687, but he certainly kept busy. He carried out experiments in alchemy and studied theology, devoting more time to these activities than he ever spent on science. He served as Master of the Royal Mint and as an MP, receiving a knighthood for his political activities rather than for his science; and he was President of the Royal Society from 1703 onwards, following the death of his bitter rival Robert Hooke. Significantly, Newton's last great scientific work, his book *Opticks*, was published just a year later, although the work it described had been completed many years before. Newton had deliberately waited until Hooke was dead so that he could publish his theory of light without any chance of Hooke replying.

One key feature of Newton's work on light is relevant to our story. It was based on the idea that light is carried by a stream of particles, like tiny bullets. The theory worked quite well at explaining phenomena such as reflection and refraction, although there was a no less successful rival theory,[3] developed by Christiaan Huygens (1629–95) in the Netherlands, that described light in terms of waves, like ripples on a pond. Newton's version held sway for more than a hundred years, until the beginning of the nineteenth century, partly because of the status of Newton himself as the acknowledged "greatest scientist who ever lived" and partly because Huygens, like

Hooke, was dead by 1704 and Newton had the last word. But then, everything changed.

Maxwell and the world of waves

In the early nineteenth century, the received wisdom about the nature of light as a stream of particles was overturned as a result of the work of two men. The first, Thomas Young (1773–1829), was an English polymath from a wealthy family. Although he trained and practised as a doctor, he had the luxury of not needing to rely on medicine for an income, so was able to devote much of his time to science, especially the nature of vision, how we perceive colours, and most notably the wave theory of light. The second was the Frenchman Augustin Fresnel (1788–1827), an engineer who worked diligently under the Napoleonic regime, but came out as a royalist when Napoleon was defeated and exiled to Elba. As a result, when Napoleon returned briefly to power (during the "Hundred Days" of 1815), Fresnel was sacked and placed under house arrest. There he developed his ideas about light before Napoleon was beaten at Waterloo and Fresnel returned to his job as an engineer.

Historically, the work of both men is of equal importance; but I shall concentrate on one key experiment, carried out by Young, because it later played a crucial part in the development of an understanding of quantum physics. It is called, for reasons that will become obvious, the "double slit experiment" or, more colloquially, "the experiment with two holes." Much later, the great American physicist Richard Feynman (1918–88) said that the experiment with two holes encapsulates the "central mystery" of quantum physics. We will see why later.

In Young's experiment, a beam of light (ideally, a single pure colour) is shone through a narrow slit in a thin sheet of card in a dark room, to produce a narrow beam. The light spreads out from the slit on the other side, and then encounters a second sheet of card, this time with two parallel slits in it. Finally, light spreading out from each of those two slits falls on a sheet of white card, which acts as a screen on the far side of the experiment. The question Young sought to answer was: "What pattern does the light make on the final screen?"

Our everyday experience tells us that if light travels like a stream of tiny particles, they should pass through the slits in straight lines and go on to hit the final screen. There should be a pile-up of particles arriving just behind each of the slits, corresponding to two stripes of light on the far screen, with the brightness fading away either side of each stripe. On the other hand, as anyone who has watched ripples spreading out from two stones dropped simultaneously into a pool of still water knows, waves spreading out from the two slits would overlap and interfere with one another, making a much more complicated pattern of light and shade on the far screen. That is exactly what Young found—the interference pattern corresponding to waves. There was no trace of the simpler pattern that would be produced by particles.

Although Young carried out these experiments and published his results in the first decade of the nineteenth century, it was not until the 1820s, and even then in no small measure thanks to the complementary work of Fresnel, that the wave theory of light began to be accepted, and it took even longer for the theory to be fully worked out. What seemed the final solution to the puzzle of the nature of light came from the investigation of something that seemed completely unrelated at the time of Young and Fresnel—electromagnetism.

Indeed, in the 1820s there was no single subject of electromagnetism, but two seemingly separate fields of study, electricity and magnetism. The person who brought the two together was Michael Faraday (1791–1867), an archetypal example of the kind of self-made man who became a symbol of success in Victorian Britain.

Unlike Young, Faraday was not born with a silver spoon in his mouth. The son of a blacksmith, he worked as a bookbinder's apprentice and then as laboratory assistant (among other things, literally a bottle-washer) at the then new Royal Institution in London, becoming a protégé of Humphry Davy (1778–1829). He was such a success that in 1825 he took over from Davy as Director of the Laboratory at the RI, and went on to become Professor of Chemistry there. It was Faraday who showed, in 1821, how an electric current could produce a magnetic field, and then, ten years later, how a moving magnet could produce an electric current. These discoveries led to the invention of the electric motor and the generator, as well as to the realization that electricity and magnetism are two facets of a single phenomenon, electromagnetism. But Faraday lacked the mathematical skill to develop a complete theory of electromagnetism, a triumph that was achieved in the 1860s by the Scottish physicist James Clerk Maxwell (1831–79).

Maxwell came from a moderately well-off Scottish family. His father owned farmland in Galloway, in the south-west corner of Scotland, where young James was brought up. His mother was forty when James was born, and had already had a daughter, Elizabeth, who had died without reaching her first birthday; James remained an only child, and his mother died when he was just eight years old. From the age of ten, he was educated in Edinburgh: first at school, staying with an aunt in the city during termtime, then from the age of sixteen at the university. He didn't complete a degree there, but moved on to Cambridge when he was nineteen, graduating from Trinity (Isaac Newton's old college) in 1854. It's a sign of how well he had done as an undergraduate that he was able to stay on at the college as a "bachelor scholar" with the intention of applying for a fellowship; however, the thought of staying at Trinity indefinitely did not appeal to him, since even in the mid-1850s Fellows of Trinity were still required to remain unmarried and (eventually) to take Holy Orders.

Over the next couple of years Maxwell refined Young's work on colour vision, showing how different mixtures of three basic colours (red, green, and blue) could fool the eye into seeing many colours (the basis of modern colour TV), and wrote an important review of Faraday's work on electromagnetism. But in 1856, not long after his father had died, Maxwell took up a post as Professor of Natural Philosophy at Marischal College in Aberdeen—where, at twenty-five, he was fifteen years younger than the youngest of his fellow professors. The most important work he carried out there was his proof that the rings of Saturn could not be solid objects but must be made up from many tiny "moonlets," each in its own orbit. The most important development in his private life was his marriage to Katherine Mary Dewar, the daughter of the College Principal. But in spite of the family connection, when Marischal College merged with King's College in Aberdeen he lost his job in the reorganization and temporarily moved back to the family home in Galloway before becoming, in

1860, Professor of Natural Philosophy and Astronomy at King's College in London.

It was while he was at King's that Maxwell's ideas about electromagnetism, which he had been puzzling over for years, came to fruition. In 1861 and 1862 he published a set of four scientific papers presenting a mathematical description of how electromagnetic waves could be transmitted. These equations naturally contained a number corresponding to the speed with which electromagnetic waves move, and to Maxwell's surprise and delight this number turned out to be exactly the speed of light, which had been measured accurately by experiments just over ten years earlier. This meant that light must be a form of electromagnetic wave—or, as Maxwell put it, "we can scarcely avoid the inference that *light consists in the transverse undulations of the same medium which is the cause of electric and magnetic phenomena*" (his emphasis). Both theory and experiment had proved, by the middle of the 1860s, that light is a wave. But the particle theory of light would make an astonishing revival less than fifty years later.

In 1864, Maxwell made a comment that is equally profound, but in a different way. He said that "scientific truth should be presented in different forms and should be regarded as equally scientific whether it appears in the robust form and vivid colouring of a physical illustration or in the tenuity and paleness of a symbolic representation." This prescient proposal accurately foreshadows the way physics developed in the 1920s. Then, as we shall see, there were two ways to describe the quantum world. One approach, pioneered by Werner Heisenberg, depended on abstract mathematical symbolism; the other, Erwin Schrödinger's brainchild, drew on the robust (and comfortingly familiar) imagery of waves. But both gave exactly the same answers to quantum puzzles; both were equally valid.

It was also in 1864 that Maxwell published his last word on electromagnetism, a paper which presented four equations that between them contain everything there is to know about electricity and magnetism, except for a few quantum phenomena. This was the greatest achievement of theoretical physics since Newton's *Principia*, and effectively marked the end of the "classical" (that is, pre-quantum and pre-relativity) period of physics. Maxwell had also achieved something else—completing the job begun by Faraday, he had shown how what had been regarded as two separate forces of nature, electricity and magnetism, could be combined into one package, electromagnetism. This was the first step towards unifying all the forces of nature in one mathematical package, a dream that drove much of Schrödinger's late work. But it was not Maxwell's own last word on physics.

In 1866, Maxwell had to resign his post at King's because of ill-health, and although still only thirty-five retired to Galloway, where he wrote a book, *Treatise on Electricity and Magnetism*, published in two volumes in 1873. But by then his health had greatly improved, and in 1871 he had been tempted out of retirement to become the first Cavendish Professor of Experimental Physics at Cambridge University and to establish and run the new Cavendish Laboratory, which opened in 1874. This became the most important centre for experimental physics in the world during the decades that followed; but Maxwell barely lived long enough to see it up and running, dying in 1879—like his mother, at the age of forty-eight. The icing on the cake of Maxwell's theory came in the following decade, when the German physicist Heinrich Hertz (1857–94) confirmed experimentally the existence of what we now call radio waves—electromagnetic waves with wavelengths much longer than those of light—which Maxwell himself had predicted.

Even his success at establishing the Cavendish Laboratory, however, does not exhaust the list of Maxwell's contributions to physics. Alongside his work on electromagetism, he had pioneered the application of statistical techniques to work out the way in which the behaviour of large numbers of atoms and molecules, speeding about and bouncing off one another and off the sides of any container they were in, explains the properties of gases, such as the way the pressure, temperature, and volume of a gas are related. The resulting kinetic theory of gases established that heat is a form of motion, and finally killed off the earlier idea that it is a kind of fluid, dubbed "caloric." As early as 1859, while still in Aberdeen, Maxwell calculated that molecules of air at a temperature of 16°C undergo more than 8 billion collisions per second, and that the average distance they travel between collisions is (in the units he used) 1/447,000 of an inch. But far more important for our story than the specific numbers that he used was the idea behind his calculations. He found a statistical law, now known as the Maxwell distribution, which did not specify the speeds of individual molecules, but specified the proportion of molecules with speeds in any particular range—the relative numbers with speeds between 14 and 15 miles per minute, between 15 and 16 miles per minute, and so on. This was the first application of a statistical law to physics—the beginning of an approach which would lead to the birth of quantum theory, and would profoundly influence Schrödinger. Maxwell developed these ideas further in the

1860s, partly through his correspondence with the Austrian physicist Ludwig Boltzmann (1844–1906), who then developed them further still.

Boltzmann and the world of statistics

Boltzmann was born in Vienna, where his father was a tax official, and like Schrödinger received his early education at home, before moving on to high school in Linz, where his father had been posted. His father died when Ludwig was fifteen, but this did not affect his education, and in 1863 he began his studies at the University of Vienna, where his teachers included Josef Loschmidt (1821–95) and Josef Stefan (1835–93). In this environment, Boltzmann became an early enthusiast for the kinetic theory (it was Stefan who introduced him to the work of Maxwell), and a firm believer in the reality of atoms at a time when the concept was still controversial. With Stefan as his supervisor, Boltzmann completed his studies for a PhD in 1866, with a dissertation on the kinetic theory. Although he was still only twenty-two, this was not quite the precocious achievement it might seem today, since in the German-speaking world at that time a PhD was a first degree, although it did involve some original work. A year later, Boltzmann became a *Privatdozent*, working as Stefan's assistant for two years before moving to Graz as Professor of Mathematical Physics in 1869.

Although based in Graz, over the next few years Boltzmann took the opportunity to visit the universities of Heidelberg and Berlin for extended periods, keeping up to date with the latest ideas in physics. The first of these visits to Berlin, in 1870, coincided with the Franco-Prussian War, which saw the birth of modern Germany; although Austria kept out of the conflict, Boltzmann cut short his visit, but returned for a longer stay when the war was over. In 1873 he returned to Vienna as Professor of Mathematics, but remained for only three years before going back to Graz to take up the post of Professor of Experimental Physics, still aged only thirty-two. The same year, 1876, he married twenty-two-year-old Henriette von Aigentler, one of the first women to be permitted to attend science lectures at an Austrian university, although she was not allowed to take a degree. The marriage produced three daughters and two sons, although the older boy, another Ludwig, would die of appendicitis when he was ten. The fourteen years following his marriage seem, apart from this loss, to have been the happiest, and certainly the most productive, period of Boltzmann's life. Unfortunately, though, he seems to have suffered from bipolar disorder, and the happiness would not last. The depression may have been triggered by the death in 1885, at the age of seventy-four, of his mother, to whom he had been close since his father's death so much earlier.

Boltzmann's interest in atoms and the kinetic theory led him to develop an explanation of the laws of thermodynamics, which had been derived by empirical studies, based on the statistical behaviour of very large numbers of particles. This became known as statistical mechanics; it was also derived, completely independently, by the American Willard Gibbs (1839–1903), but his ideas did not cross the Atlantic at that time.[4] The statistical approach provides insight not only into the behaviour of particles, but also into the behaviour of radiation, with important implications for quantum theory that I will describe later. But for now the implications for particles will give a good idea of the significance of Boltzmann's work.

The simplest way to get a handle on statistical mechanics is in terms of the famous second law of thermodynamics, which says that the amount of disorder in a system that is left to its own devices (a closed system) always increases. In everyday language, things wear out—if I drop a wine glass it breaks, but shards never spontaneously combine to make a glass. The act of making a wine glass does not violate the second law because it does not happen in a closed system—it involves an input of energy from outside.

But is it literally true that shards of glass can *never* re-assemble themselves? This is one of the puzzles that Boltzmann addressed. The problem can be expressed in even simpler terms by imagining a sealed box full of gas. Everyday experience tells us that the gas fills the box uniformly; it doesn't all gather at one end. Indeed, if we have a box with a sliding partition, and put gas in one half of the box with the partition closed, we can be sure that when the partition is slid away the gas will spread out to fill the box. It will never move back into one half of the box, giving us a chance to slide the partition back and keep it there.

Or will it? The puzzle is that, according to Newton's laws of mechanics, every collision between atoms is reversible. If we made a movie showing the gas spreading out to fill the box, then ran the movie backwards, it might look bizarre, but there would be nothing going on in the time-reversed version of events that conflicted with Newton's laws. In 1890, the French physicist Henri Poincaré (1854–1912) established that in a box of gas like this, every possible arrangement of the atoms in the box must occur sooner or later.

Boltzmann's resolution to the puzzle was to point out that although there is nothing in Newton's laws to prevent all the gas gathering in one half of the box, the statistical likelihood of this happening is very, very small. If you wait long enough the gas will all gather in one end of the box; if you wait longer still, the wine glass will reconstruct itself. But the time required for these kinds of events to have a high probability of occurring are mind-bogglingly large—far longer than what we now estimate to be the age of the Universe. In the everyday world things behave the way they do, in thermodynamic terms, because they are overwhelmingly the most likely things to happen, statistically speaking; not because it is literally impossible for things to be otherwise. As Boltzmann emphasized in a paper published in the science journal *Nature* in 1894, the second "law" of thermodynamics is actually only a statement of probability. In a world governed by statistical rules, never say never.

By the time that paper was published, though, Boltzmann had moved on again, and his life was heading for an unhappy ending. In 1890 he had become Professor of Theoretical Physics in Munich, but in 1893 he had the irresistible opportunity to move back to Vienna as Stefan's successor in the equivalent post there. Unfortunately, this would bring him into direct conflict with Ernst Mach (1838–1916), who had returned to Vienna from Prague (where he had been Professor of Experimental Physics since 1867) to become Professor of History and Theory of the Inductive Sciences in 1895. Mach was a fine experimental physicist who studied air flow over moving objects, and whose name is immortalized in the "Mach number" used to indicate the speed of an aircraft relative to the speed of sound. But his philosophy was more contentious, and it was on this ground that Mach and Boltzmann would clash. Mach subscribed to the positivist view that only things we experience directly with our senses are real (indeed, he is known as the "father of logical positivism"). This made him the leading opponent of the idea of atoms, which he regarded as merely heuristic devices, rather in the way that in the seventeenth century the Catholic Church said that it was acceptable for Galileo to use the idea of planets orbiting the Sun as a heuristic device to make calculations easier, but it was not acceptable for him to teach that planets really do orbit the Sun.

It was unfortunate that at a time when the idea of atoms had already been accepted by almost all chemists and by the great majority of physicists, Boltzmann was based in what was effectively the last stronghold of opposition to the idea. It was also unfortunate that Boltzmann and Mach did not get on well personally. Finding these tensions particularly hard to take in view of his mental health problems, in 1900 Boltzmann moved to the University of Leipzig. There too he encountered professional disagreement, this time with Wilhelm Ostwald (1853–1932), a leading positivist who strongly denied the reality of atoms and maintained this position until 1908. Even though Boltzmann was on good terms with Ostwald personally, as he had not been with Mach, he was unhappy in Leipzig, troubled by failing eyesight and a developing fear of lecturing, worrying that his mind was losing its sharp edge and he might start talking nonsense. He also suffered badly from asthma. Fortunately, in 1902, after Mach had retired because of ill-health, Boltzmann was able to return to Vienna as an ordinary professor with few duties. He made a second visit to the United States in 1904, accompanied by his son Arthur, and yet another trip across the Atlantic in 1905, this time alone, to give lectures at the Berkeley campus of the University of California.

Even though Mach was no longer on the scene, in his darker moments Boltzmann still felt, entirely erroneously, that his ideas had fallen on stony ground and were not being taken seriously. In 1898, in the middle of his difficult time alongside Mach in Vienna, Boltzmann had written in one of his scientific papers that he was publishing his calculations in the hope "that, when the theory of gases is again revived, not too much will have to be rediscovered," seemingly unaware of how well his ideas had been received in the English-speaking world. This misperception was undoubtedly a factor in what happened next. Colleagues knew of Boltzmann's mood swings, although bipolar disorder was not understood at the time, and some were aware that in his blackest moments he had previously attempted suicide. As he entered his sixties, he had to give up much of his teaching because of these problems. So it cannot have come as too much of a shock to them when, in 1906, he hanged himself while on holiday in Italy. But the news did come as a shock to Erwin Schrödinger, then a young student who was about to begin his studies at the University of Vienna, and who had hoped to learn directly from one of his scientific heroes.

Chapter Three

Twentieth-Century Man

When Erwin Schrödinger began his studies at the University of Vienna in the autumn of 1906, Boltzmann had just died, and the new professor, Friedrich (Fritz) Hasenöhrl (1874–1915), would not be appointed until the following year, leaving physics teaching at the university in limbo for eighteen months. But Hasenöhrl was the ideal man for the job, having studied under both Stefan and Boltzmann, and taught at the Technical High School in Vienna. In 1904, while studying the theoretical relationship between mass and energy, he had come close to producing the special theory of relativity a year before Albert Einstein made his famous breakthrough. When he was finally appointed to Boltzmann's old chair, Hasenöhrl hit the ground running with a tour de force inaugural lecture summing up Boltzmann's work on statistical mechanics; Schrödinger, starved of real physics for over a year, was hooked, and immediately resolved to make it his life's work to follow where Boltzmann had led. Young and energetic, a brilliant lecturer and up to date on the latest ideas in at least some areas of physics, Hasenöhrl was a far better inspiration for Schrödinger than the disillusioned old man Boltzmann had become would have been.

Student life

By the time he heard Hasenöhrl lecture, Schrödinger already had a reputation as an outstanding student—indeed, he had brought that reputation with him from the *Gymnasium*, where he graduated first in his class, and was sometimes referred to by his fellow undergraduates as "the" Schrödinger. Although Schrödinger had few close friends among his contemporaries at the time, he was well liked, and could always be relied on to help out weaker students with their maths and physics. His closest friend was a botanist, Franz Frimmel, in spite of the fact that Frimmel had deep religious convictions. He also became firm friends with one of the slightly older members of the university, Fritz Kohlrausch, who completed his first degree while Schrödinger was halfway through his own course, and stayed on as an experimental physicist. This friendship lasted for life, and grew to include the two scientists' families.

But the major influence on Schrödinger's life from 1907 to 1910 was Hasenöhrl, who lectured to the students five days a week. Schrödinger later said that the only person who had a bigger influence on his life was his own father. And the influence extended outside the lecture room: Hasenöhrl, a winter sports enthusiast, organized group expeditions with students and, along with his young wife, welcomed them into his home.

It was just as well that Hasenöhrl was such an enthusiastic, friendly, and able lecturer, since the teaching facilities provided for him were a disgrace. Although the impressive main university building had been completed in 1884, the physicists were still stuck in "temporary" accommodation acquired in 1875. Students had to sit on ordinary chairs in the lecture room, with their notebooks on their laps; the floor was cracked and polluted with mercury from the room's time as a laboratory; and the building was so badly constructed, according to Schrödinger's contemporaries, that the walls would shake in a strong wind. In these conditions Schrödinger sat through not only the inspiring lectures given by Hasenöhrl, but also a lot more routine material, ranging from chemistry to calculus—and including one course which, although seeming nothing special at the time, may literally have saved his life: meteorology.

In the wider world, there were stirrings of change. Following demonstrations and marches in support of the right to vote, all adult male subjects were allowed to cast their ballots in the Austrian election held in May 1907. But this was little more than window-dressing; although the body they elected was sometimes referred to as the "people's parliament," in practice power remained in the hands of the Emperor and his advisers, and the Empire continued on its corrupt, ramshackle, and bureaucratic way. The following year Schrödinger's personal life was stirred, when he had a brief but passionate affair with a girl called Ella Kolbe. Although Erwin was still living in the family apartment in the centre of Vienna, he was able to meet Ella at an apartment near the university where one of his fellow students, Jakob Salpeter, lived. He saw a lot of Salpeter over the next few years while they worked in the same laboratory on experiments to complete the research requirement for their degrees.

The title of Schrödinger's dissertation was "On the Conduction of Electricity on the Surface of Insulators in Moist Air," and it is just about as dull as the title suggests. Schrödinger did little more than the bare minimum that was required in order to obtain his degree, and the only good thing to be said about the project is that it gave him experience of practical laboratory work that would come in useful later. The rest of his work, though, was of his usual high standard, and having passed out first in his class Schrödinger was duly awarded his Doctor of Philosophy degree (about equivalent to a modern MSc) in June 1910. The next step, after a summer break, was military training.

Life beyond the lab

In the Austro-Hungarian Empire at that time, all able-bodied young men were required to do three years of military training. At least, hoi polloi had to serve for three years. The educated upper classes were allowed instead to "volunteer" for officer training, which took only one year. If they chose, this experience of military life could be pretty easy, since the cadets were not even required to pass the final examination which would qualify them to be commissioned in the reserve. The stratification of Austrian society was reflected in the army, where the cavalry were regarded as the élite, the artillery were of lower social standing (although the fortress artillery held themselves to be superior to the field artillery), and everybody looked down on the poor bloody infantry. In October 1908, Schrödinger enrolled in the fortress artillery, which accurately reflects the position of his family in Viennese society.

Thanks to his service record, we know that Schrödinger was 167.5 centimetres (5 feet 6 inches) tall, with blue/green eyes and fair hair. Unlike many of his fellow volunteers, he took his duties seriously, although in fact they were not onerous. After the first two months living in barracks, the young men were allowed to find their own lodgings nearby, where they were responsible for their own expenses. Leave came as early as Christmas, at which point Schrödinger went skiing with a fellow physicist, Hans Thirring (1888–1976), who was then in his final year at the university. This casually taken holiday had an almost immediate effect on Schrödinger's career. Thirring broke his foot, and as a result was excused military service, so he was available to become Hasenöhrl assistant when the post became vacant; other things being equal, the job would surely have gone to Schrödinger, the best student in his year.

Thirring went on to become a leading physicist and a pacifist, active in the Socialist Party of Austria. He is best known for his work on the general theory of relativity; but some of our knowledge of Schrödinger's student days comes from Thirring's reminiscences. Schrödinger went on to complete his military training without further incident, and was duly commissioned in the reserve with the cadet rank of *Fähnrich*, just below that of lieutenant. Free to return to the University of Vienna, instead of working on theory as he would have done alongside Hasenöhrl, in 1912 he became an assistant to Franz Exner (1849–1926), as an experimenter with responsibility for the practical classes of the first-year physics students.

In order to become a *Privatdozent*, the first step on the ladder towards a professorship, Schrödinger had to carry out original scientific research. Significantly, he chose problems in theoretical physics, not experimental work, although he did work on topics relevant to Exner's studies. Competition for the few places available for *Privatdozenten* was fierce, but Schrödinger never underestimated his own ability, and as usual his confidence was justified. He earned his promotion with a study of the nature of magnetism which, though correctly worked out, started (we now know) from a false assumption, and a study of the way solids melt which was groping towards an understanding of the way atoms and molecules interact in solids and liquids. He could hardly have achieved more at the time, since the work was carried out in 1912, exactly when William Bragg (1862–1942) and his son Lawrence (1890–1971) in England, and Max von Laue (1879–1960) in Germany, were beginning to study the crystalline structure of solids using X-rays.

Hasenöhrl, on behalf of the examining committee, gave Schrödinger a glowing report—"In the opinion of the committee, all the works of Schrödinger demonstrate a very well founded and broad scholarship and a significant original talent"—and in spite of the opinion of one committee member that the appointment was premature on the grounds of Schrödinger's youth, after months of further formalities including an oral examination and the need to have the appointment confirmed by the Ministry for Culture and Instruction, Schrödinger was appointed as a *Privatdozent* in the University of Vienna in January 1914, at the age of twenty-six.

But while all this had been going on, Schrödinger had been distracted by love, and up until the middle of 1913 had been seriously considering abandoning an ill-paid academic career in order to join his father's business and make enough money to support a wife and family. The notion horrified Rudolf, who had had to give up his own academic ambitions for business, and delighted in Erwin's progress; but in the end it came to nothing.

The object of Erwin's affections at this time was typical of the girls he would later be involved with, and in the future his amorous adventures would have a profound influence on his academic career. She was Felicie Krauss, the daughter of Karl and Johanna Krauss, family friends who happened to be wealthy, strict Catholics, and socially a step above the Schrödingers, with pretensions to (minor) aristocracy (they were entitled to the prefix "von" before their surname). Felicie was nine years younger than Erwin, and as a child he had often, to his disgust, had to look after the little girl when the two families got together. When her father died in 1911, Felicie was not quite fifteen, and Erwin's feelings were now quite the reverse—he was happy to be with her on every possible occasion, although what was possible was strictly limited by the conventions of polite society. Johanna Krauss was alarmed by the developing relationship between Felicie and Erwin, who as an impoverished, free-thinking (perhaps even atheist) academic was in her eyes quite unsuitable husband material. She forbad the couple to meet more than once a month—which, of course, increased their ardour until they became informally engaged.

It was at this point that Erwin asked his father about giving up physics to join the linoleum business. Rudolf was adamant that Erwin should not make the same sacrifices that he had had to make, and Johanna Krauss was adamant that the marriage would never take place. At her mother's behest,

Felicie told Erwin that it was all over in the summer of 1913, when he was halfway through the process of becoming a *Privatdozent*. But Felicie, who later married a lieutenant in the Austrian army from a similar social background to her own, remained friends with Erwin, and later with his wife, Anny. Schrödinger threw himself into his work, and in March 1914 produced his first really significant scientific paper, developing some of Boltzmann's ideas and improving the mathematical description of the interactions between atoms in molecules. His relationship with Felicie may have been short-lived but, in one way, it was formative. Erwin never lost his fascination with young girls on the brink of adolescence.

Despite his personal dilemmas, in 1912 and 1913 Schrödinger did not neglect his experimental duties as Exner's assistant. And those duties brought him into contact with the girl who would replace Felicie in his affections and eventually become his wife.

One of the research interests of Exner's group was the study of atmospheric electricity. This involved measuring the electrical conductivity of the air at various locations and different times, and measuring the background radiation which made electrically charged instruments (electroscopes) gradually lose their charge. The radiation turned out to come from two sources. One was naturally occurring radioactive material, such as radium in the rocks; the other was radiation penetrating the Earth's atmosphere from outer space. This latter radiation, now known as cosmic rays, was first discovered by Victor Hess (1883–1964), who worked at the Institute of Radium Research of the Viennese Academy of Sciences. He made the discovery on high-altitude (and highly dangerous) balloon flights in 1912; it earned him the Nobel

Prize in Physics, though not until 1936, nearly twenty-five years later.

Schrödinger's contribution was more down to Earth. In 1910, Fritz Kohlrausch had made some measurements of atmospheric electricity at the lakeside resort of Seeham, on the Mattsee. In 1913, Exner decided that it would be a good idea to repeat the measurements to see if there had been any change, and gave Schrödinger the far from onerous task of spending the summer, from late July to early September, on the job. This was just after the split with Felicie.

The work was dull rather than demanding, and produced no great scientific discoveries. Schrödinger was able to take full advantage of the opportunities for hiking and swimming, even though it was a wet summer, and to enjoy the company of his friends the Kohlrausches when they brought their children for a holiday at the lake. They also brought with them a teenage girl from Salzburg, Annemarie (Anny) Bertel, to look after the children. She had been born on the last day of 1896, so in the summer of 1913 she was only sixteen, and still in pigtails. Fifty years later, in an interview for the Archives for the History of Quantum Physics, she recalled being impressed by the "very good looking" young scientist, and he clearly noticed her, but nothing significant passed between them at the time.

Much more significant was the event that hastened Schrödinger's return to Vienna from Seeham early in September. A splendid new physics building had at last been opened in the spring of 1913, and that autumn it was the setting for a major scientific conference, known as the Congress of Vienna, involving more than seven thousand scientists, including the rising star Albert Einstein, and endowed with all the pomp and glamour of the last days of the Empire, up to and including an Imperial Reception. Apart from Einstein, who discussed the need to modify Isaac Newton's theory of gravity, the speaker who made the biggest impression on Schrödinger was Max von Laue, who described his work on X-ray crystallography. In 1914, the first course Schrödinger gave after he became a *Privatdozent* was entitled "Interference Phenomena of X-rays"; it was only later that he got his teeth into the problem of gravitation. But he had barely got into his stride as a lecturer when the First World War intervened. Vienna—and Schrödinger—would never be the same again.

War service on the Italian Front

When Archduke Franz Ferdinand, the heir to the imperial throne of Austria-Hungary, was assassinated in Sarajevo on 28 June 1914, the old Emperor, Franz Josef, shed only crocodile tears. The assassination was treated equally calmly by most of Viennese society. Franz Ferdinand was widely regarded (not least by his uncle, the Emperor) as unsuitable for the top job, as indicated by, among other things, his decision to marry for love. His wife, Sophie, a former lady-in-waiting, was so much his social inferior that the marriage had only been permitted on a morganatic basis—that is, with the stipulation that any children would not inherit Franz Ferdinand's titles and right of succession. With such a dangerous free-thinker out of the way, the heir apparent was Franz Josef's great-nephew Charles, who was much more of a traditionalist and would probably have made an excellent nineteenth-century Emperor.

The trouble was, it was no longer the nineteenth century, and Charles would never get a real chance to prove his mettle.

The assassination of Franz Ferdinand provoked Austria into attacking Serbia (after all, even the assassination of an unpopular heir could not go unpunished), and the obligations this triggered in a complex web of international treaties—where A promised to attack B if B attacked C, but D had promised to defend B from attack by A, and so on—resulted in the First World War. When all the treaty obligations had been invoked, the Central Powers of Germany and Austria-Hungary were left facing (and more or less surrounded by) the Triple Entente of Britain, France, and Russia, soon joined by Italy, and lesser allies—including, of course, Serbia. Reserve artillery officer Erwin Schrödinger received his call-up papers on the last day of July and went with his father to buy two pistols, neither of which he ever had to use. He also found time before leaving Vienna to send a gift to Anny Bertel—a book of essays by the Austrian writer and critic (and author of *Bambi*) Felix Salten.

Schrödinger was sent to a fortified artillery position near the Italian border, high in the mountains overlooking the Venetian plain. This was as good a place as any to be posted in the summer of 1914, far removed from the fierce fighting on the Russian Front, where the Austrians lost a quarter of a million killed or wounded and a hundred thousand prisoners in the first three weeks. By contrast, with Italy not yet involved in the war, things were so quiet in the mountains that Schrödinger was even able to carry out some scientific work, completing calculations based on experiments he had carried out in Vienna. After all, he needed nothing for this work except paper, a pencil, and a few books—unlike many mathematical physicists today, who would be lost without a computer. So on 27 October 1914 Schrödinger was able to send a short paper on the pressure exerted by gas bubbles to the journal *Annalen der Physik*.

Before the winter, Schrödinger was posted to a fortress in the South Tyrol, dominating the entrance to the Brenner Pass. He spent the winter of 1914–15 there, enjoying the beautiful mountain scenery while the conflict on the Western Front settled into the grim trench warfare for which the First World War is best remembered by Britain and the other Western participants. His next posting was to the equally peaceful, but less scenic, garrison town of Komárom, between Vienna and Budapest. There, Schrödinger wrote a paper about the behaviour of small particles being jostled in a fluid (gas or liquid) by the impact of molecules of the fluid. This is known as Brownian motion, after the Scottish physicist Robert Brown (1773–1858), who studied it in the 1820s. In 1905, Albert Einstein had proved that this erratic jittering can be explained statistically as caused by the constant but uneven bombardment that particles such as pollen grains receive from atoms and molecules, and thereby provided compelling evidence for the reality of atoms—just too late for this to be much comfort to Boltzmann.[1]

In a quite separate investigation, culminating in 1912, the American Robert Andrews Millikan (1868–1953)—who also, incidentally, coined the term "cosmic rays"—had managed to measure the charge on the electron by monitoring the way tiny electrically charged droplets of water or oil drift in an electric field. These droplets are small enough to be affected by Brownian motion, and Schrödinger analysed statistically the importance of these effects in Millikan-type experiments. Nothing dramatic came out of the study, but it is important in the context of Schrödinger's career because it was his first published foray into statistics, which would later loom large in his work.

By the time this paper was published, the war, and Schrödinger, had both moved on. Italy was persuaded to join the Triple Entente with promises of large chunks of Austria, and declared war on 23 May 1915. As part of the Austrian response, Schrödinger's unit was moved to Oreia Drega, near Görz (later Gorizia), north-west of Trieste. This would be the scene of fierce fighting at various stages in the war (immortalized by Ernest Hemingway in *A Farewell to Arms*), but the artillery were well back from the front line and engaged the enemy at long range, sometimes coming under heavy fire but suffering relatively few casualties. Schrödinger was even able to visit Görz when off duty, relaxing in one or another of the city's coffee houses. Judging from his diaries, the worst problem he had to face during the summer of 1915 was boredom; but in September he received some books and copies of scientific journals, and the diary entries stop as, presumably, he began a bout of work and study. Soon, he also began the most active phase of his military career, for in October and November 1915 Schrödinger was acting commander of the battery, in charge during several fierce engagements and behaving with sufficient distinction to receive a military citation.

The fighting eased up for the winter, but in May 1916, just after Schrödinger had been promoted to *Oberleutnant*, it flared up again in the Görz sector, with losses amounting to more than a hundred thousand on the Austrian side and more than a quarter of a million on the Italian. But by then Schrödinger had been posted to command a battery in the mountains north of Trieste in what he later called "an extremely boring but beautiful spot." If you had to be a soldier, it was better to be in the artillery than in the trenches—a point brought home by the death of Fritz Hasenöhrl leading an infantry charge in the Tyrolean sector of the front.

Things began to change again at the end of 1916. On 21 November Franz Josef died; his successor, Charles, made strenuous efforts to conclude a peace with the Entente, but met stiff resistance from the Italians, who insisted on having all the territory they had been promised by their allies. Also, and more crucially, the military rulers of Germany, by now very much the senior partner of the Central Powers, refused to let the Austrians negotiate an independent peace. Meanwhile the war continued quietly for Schrödinger. He even enjoyed a visit from Anny Bertel, who had turned twenty at the end of 1916. According to Schrödinger's notebooks, which rather ungallantly indicate all his lovers (albeit in code), they did not become intimate at this time, but clearly something was in the air. Anny was the only one of his female friends who visited Schrödinger while he was serving on the Italian Front. But soon his friends in Vienna would not have to travel far to see him.

Back to Vienna

The weather—or rather, meteorology—came to Schrödinger's rescue in the spring of 1917. He was posted back to Vienna to teach a course in meteorology for antiaircraft gunnery officers, combining this with teaching an introductory practical physics course at the university. It is possible that, following the death of Hasenöhrl and the succession of

Charles, the authorities in Vienna wanted to ensure that there was at least one good physicist around to pick up the threads at the university after the war. Or he may just have been lucky.

Whatever the reasons for the posting, it gave Schrödinger the chance to recommence research, and to begin publishing scientific papers again. The most significant of his 1917 papers, in the light of later developments, was his first application of quantum theory (of which much more in the next chapter) to a scientific problem. The problem had to do with the heat capacity of solids—a measure of how much energy is required to increase the temperature of a certain amount of material by a certain number of degrees. This is related to the way molecules vibrate, which in turn depends on their quantum properties. Schrödinger's contribution followed a line of enquiry, rooted in thermodynamics, which had also been pursued by Max Planck and Albert Einstein. In one study he investigated the random fluctuations in the rate at which samples of radioactive material decay (related to the concept of the half-life, the time in which exactly half of the atomic nuclei in a sample will decay).

Schrödinger also became interested in the general theory of relativity, which Einstein published in 1916; news of the breakthrough reached Schrödinger while he was still serving on the Italian Front. Einstein's description of gravity, and the relationship between gravity and matter, in terms of curved spacetime completely changed the way many physicists thought about the fundamental properties of the Universe, Schrödinger among them. He wrote two papers in 1917 on the implications of the general theory: the first investigated the description of energy provided by Einstein's equations; the second addressed the nature of the Universe itself. He found a solution to Einstein's equations which describes a universe completely devoid of matter but in which empty space has tension, like a stretched spring; this is clearly not a description of our Universe, but it shows both the power of Einstein's theory and the breadth of Schrödinger's interests.

Exciting as these investigations were, they were carried out against a background of increasing hardship as the war ground towards its close. The entry of the United States into the conflict against Germany in April 1917 made the ultimate outcome inevitable, although the USA did not declare war on Austria-Hungary until the end of the year. Long before then, the Allied blockade of the Central Powers had brought both Austria's economy and its army to their knees. When the bread ration was cut from 200 grams a day to 160 grams in January 1918, munitions workers went on strike, and seven divisions of half-starved soldiers (those in the front line received only 200 grams of meat a week, others half that) were brought back from the front to restore order. In Vienna, those who could afford the black market did not go hungry, but had to sacrifice their possessions in order to eat. Closer to home, things were also going from bad to worse. Schrödinger's mother underwent an operation for cancer of the breast in 1917 and Erwin himself was ill in 1918 with what seems to have been tuberculosis. The family business was ruined by the lack of raw materials. Schrödinger stayed on the army payroll until the end of 1918, but just when he lost that income, and in spite of—in fact, because of—the end of the war, the situation in Vienna became even worse.

The aftermath

Even after the Armistice of 11 November 1918, the victorious powers maintained their blockade of Austria-Hungary, and watched through the winter of 1918–19 while the Empire fell apart. This constituted one of the worst war crimes of the twentieth (or indeed any) century, but remains little known today, because, of course, history is usually written by the victors. Food supplies no longer came in from Hungary, coal no longer came in from Czechoslovakia, Italian troops (themselves reasonably but not generously well fed) occupied Vienna; so desperate did the situation become that Germany, even though itself blockaded, sent some food. Italy and Switzerland also provided some aid; the criminals were the intransigent British and French. The Emperor was deposed, although hardly anyone noticed, and Austria became a republic. As a last resort, with the empire now lost, the government tried to make the country part of Germany, but the French vetoed any such notion. On the rare occasions when food was available, mobs of women would rush the supplies, and would have to be controlled by mounted police; in one widely reported incident, when a police horse fell in the crush it was sliced up and the meat carried away in a matter of minutes.

Schrödinger was trapped in this mess. He had been offered a job as a physics lecturer in Czernowitz—but Czernowitz was now part of Romania, and the Romanian government forbad the appointment of a foreigner to such a post. In the midst of all this turmoil, he found consolation in philosophy, flinging himself into a study of Wittgenstein, Schopenhauer, and through Schopenhauer Eastern, and in particular Indian, philosophy. He was particularly fascinated by the Vedanta, a school of Hindu philosophy which teaches that there is only one reality. These philosophical ideas would strongly influence Schrödinger's later thinking about quantum physics.

The situation in Vienna began to improve after Herbert Hoover, then head of the American Relief Administration and later to become US President, visited the city in January 1919. Alarmed by the prospect of a Communist revolution being triggered by the appalling conditions, he arranged for the first American aid to be sent to Vienna, and the blockade of Austria was lifted at last on 22 March—although it continued to be applied to Germany, where the kind of revolution Hoover feared very nearly did happen. As the threat of starvation receded, other problems accumulated. Schrödinger's income from the university was modest (far too small to support a family), but he spent as much time as he could there, escaping from the mounting troubles in the family apartment, where his father was increasingly ill with hypertension and atherosclerosis, and where limited gas supplies meant that the rooms were cold and dark. The family depended on income from Rudolf Schrödinger's investments, but as the cost of living began to increase this became less and less adequate.

The metaphorical light in this gloom came from Anny. Schrödinger was able to visit her in Salzburg, and in the autumn of 1919 they became engaged; they also, according to his notebooks, became lovers in the physical sense at this time. Anny moved to Vienna, and obtained a well-paid job as a secretary—to Erwin's embarrassment, her monthly income was more than he earned in a year. Anny's employer was a Bauer, Friedrich (known as Fritz), but not a close relative of

Schrödinger's maternal grandfather. Fritz Bauer was the Director of the Phoenix Insurance Company, and lived in a grand house in the suburbs of Vienna. On one occasion in 1920 Anny took Erwin, as her fiancé, to tea here to meet the Bauer family. Fritz's thirteen-year-old daughter Johanna, known as Hansi, later remembered Schrödinger as very stiff and uncomfortable, no doubt because the surroundings emphasized his own limited prospects.[2] The light and the gloom collided even more dramatically on Christmas Eve 1919, when Anny was staying with her family in Salzburg. That evening, a basket of presents arrived from Anny; barely an hour earlier Rudolf Schrödinger, sitting peacefully in his chair, had died.

Schrödinger would later come to think that it was just as well his father had not lived to see 1920, when runaway inflation wiped out all the family's savings. But developments in Schrödinger's life moved almost as fast as the inflation that crippled the city. In January 1920 he was offered a promotion to assistant professor in Vienna, but the salary would not have been sufficient to support a wife, and, although eager to marry, Schrödinger had no intention of living off Anny's income. But he was also offered a similar post in Jena, in Germany, with a salary sufficient to allow the marriage to take place. He duly accepted, and the couple were married. In fact they were married twice—first in a Catholic ceremony on 24 March 1920, then in an Evangelical church on 6 April. She was twenty-three, and he was thirty-two. An earlier biographer, Walter Moore, has described the union in terms which are hard to beat: "She entered the marriage with the hope and expectation that it would be a true union of minds and bodies, in which she would achieve happiness through boundless submission to her brilliant and beautiful lover. These illusions may have lasted for at least a year." Nevertheless, the marriage lasted for life. As Anny herself later told Hans Thirring, "It would be easier to live with a canary bird than with a racehorse, but I prefer the racehorse."

The racehorse had not been scientifically idle in the three difficult years leading up to his marriage. One piece of work was particularly relevant to the later development of his career, and was also his last significant piece of experimental research. In the second decade of the twentieth century, there was a great deal of puzzlement about the nature of light. By the end of the nineteenth century, the work of physicists such as Thomas Young and James Clerk Maxwell seemed to have established that light was a wave. But within the first five years of the new century, the work of Max Planck and, in particular, Albert Einstein had revived the idea of light as a stream of particles. (More of this in the next chapter.) Schrödinger decided to carry out an experimental test to decide between the rival wave and particle models of the behaviour of light. His work was essentially a refinement of the double slit experiment, using a very fine electrically heated wire as the light source and studying tiny interference fringes with a microscope. The results were exactly in line with what was expected of the wave model, reinforcing Schrödinger's faith in the world view of classical physics.

But the most important work that Schrödinger carried out in the last years of the century's second decade concerned a quite different topic—the theory of colour vision. The fruits of his labours appeared in a series of scientific papers published in 1920, describing and quantifying the perception of hue, brightness, and saturation of colours, and how a change in one of these properties can affect the perception of the other two. This led to Schrödinger being asked to write the article "The Visual Sensations" for a major encyclopedia, the *Lehrbuch der Physik*; this 104-page essay, published in 1926, became the standard reference on the subject. He returned briefly to colour theory in the mid-1920s, applying his ideas to the practical problem of comparing the colours of stars with one another, an important task in determining their temperatures. But by then his main line of research was leading him not out into the Universe but in and down, to the world of atoms. He had also moved on—from Jena to Zürich, via Stuttgart and Breslau.

The peripatetic professor

Schrödinger settled in well in Jena, where he arrived with Anny in April 1920, but the post there was only a temporary one, and when he was offered a permanent associate professorship in Stuttgart he had no hesitation in moving there in October of the same year. While there, he had time to do some work on the theory of electron orbits. But by this time, financial security was becoming a paramount concern. Roaring inflation in Germany meant that Schrödinger was barely able to support himself and Anny, and had no reserves with which to help his ailing mother when her own father became so desperate for income that he had to turn her out of the large apartment in Vienna and rent it out. Georgie's family helped her to settle, as far as she could, in smaller premises, and Anny helped during her final illness. She died there, of cancer, in September 1921. The image of his widowed mother dying in such reduced circumstances, and the fear that something similar might happen to Anny, haunted Schrödinger for the rest of his life, and strongly influenced his decisions about career moves. Financial security would always come first.

In these circumstances it is no surprise that Schrödinger accepted an appointment as Professor of Theoretical Physics in Breslau in the spring of 1921, even though Breslau (now Wrocław) was uncomfortably close to the Polish border and a hotbed of Communist sympathizers—a real concern given the civil war then raging in Russia. But after some eighteen months as a peripatetic professor, and less than six months in Breslau, Schrödinger was saved by an offer he couldn't refuse from a safer haven in Switzerland.

The University of Zürich, having managed without a professor of theoretical physics during the war years, was looking, slowly, for a new one. The laborious process of finding the right man for the job, involving interminable committee meetings, began at the end of 1919 and dragged on for more than a year. One reason for the lack of urgency was that with the help of junior lecturers things were ticking over more or less OK, and the university (or rather, the cantonal government) was saving money by not appointing a professor. Then, in March 1921, Paul Epstein, who had been one of the lecturers helping to fill the gap in Zürich, left to take up a post in Leiden. This concentrated the minds of the committee, and, finding that they could not afford their first choice, Max von Laue (who had already held the post once, briefly, before the war), they offered the post to Schrödinger on the basis of his all-round ability and publications across a wide range of topics. The appointment would be for six years, commencing in October 1921, at an annual salary of 14,000 Swiss francs—the top end of the range for such a post.

Although Switzerland had suffered in the European post-war depression, it had not experienced runaway inflation and the salary was more than adequate (nor was there any risk of Communist revolution in Switzerland). Schrödinger wrote to accept the offer on 16 September, and moved to Zürich soon after his mother's funeral, making only a brief detour to wind up his affairs in Breslau.

It would be in Zürich that Schrödinger would make his major contribution to physics, during what became known as the second quantum revolution; but to put that work in perspective, we need to take a step back to look at the first quantum revolution, triggered by the work of Max Planck at the end of the 1890s.

Chapter Four

The First Quantum Revolution

The quantum revolution began just as the nineteenth century gave way to the twentieth. In December 1900, the German physicist Max Planck (1858–1947) announced that he had solved a fundamental puzzle about the nature of light and other electromagnetic radiation, using statistical techniques that Maxwell would have appreciated. But the breakthrough came at a price—it involved treating light as if it came in discrete packets, what later came to be known as quanta. To say that Planck was uncomfortable with the idea would be an understatement. He believed that light must be a wave, but that it could only be absorbed or radiated by atoms in definite amounts. Nevertheless, the equations were telling the truth. It would be left to Albert Einstein (1879–1955) to establish the reality of light quanta (also known as photons)—the work for which he received his Nobel Prize.

The puzzle Planck solved concerned the nature of what is usually known today as black body radiation, but which in the

German-speaking world of the 1890s went by the more accurate, but less dramatic, term of cavity radiation.

When black bodies are bright

To a physicist, a black body is an object that absorbs all the electromagnetic radiation that falls on it, including light. Black body radiation is the kind of radiation that would be emitted by such an object if it were hot. Cavities come into the story because the best way to simulate a black body in the laboratory is to have a hollow container, like a large insulated box, with a tiny hole in one of its walls. If you shine radiation into the hole, it goes into the box and bounces around inside, with very little escaping. But as the radiation fills the box, it will get hotter, and now if you stop shining anything into the hole radiation will escape from the cavity through that hole: this is cavity radiation. It turns out that the exact nature of the radiation being emitted (its spectrum) depends only on the temperature inside the cavity, and not on what the box is made of. This has important practical applications—for example, we know the temperature at the surface of the Sun because the radiation we receive from the Sun corresponds to cavity (black body) radiation with a temperature of just under 6,000°C, and we knew this before we knew the composition of the Sun. As this example shows, a black body can be bright—and it can be yellow, or red, or any other colour. But what puzzled Planck and his colleagues in the late nineteenth century was that according to the laws of physics they knew at that time, it should have been even brighter—indeed, infinitely bright.

Cavity radiation was first studied by the German physicist Robert Kirchhoff (1824–87) in the 1850s. The key feature of this radiation is that for any specific temperature its spectrum has a peak, corresponding to the maximum energy being radiated, at a specific wavelength, with less energy being radiated at both shorter wavelengths and longer wavelengths. A graph of the amount of energy radiated by a black body at different wavelengths rises smoothly from lower energies at shorter wavelengths to a peak at some intermediate wavelength, then slides down smoothly again towards lower energies at longer wavelengths. The wavelength of this peak depends only on the temperature of the object. In the visible spectrum, wavelength corresponds to colour, and the peak shifts to shorter wavelengths as the temperature of the object increases. This is why a lump of iron, which radiates almost like a black body, glows red when it is relatively cool, yellow when it is hotter, and blue-white when it is hotter still. But that is not what classical electromagnetic theory—the theory of Maxwell's wave equations—predicted in the second half of the nineteenth century.

If electromagnetic waves are treated mathematically in the same way as other kinds of wave, such as the waves corresponding to notes played on a violin string, the equations tell us that it is easier to radiate energy at shorter wavelengths. Indeed, the amount of energy radiated ought to be inversely proportional to the wavelength. A black body—any black body made of any material—ought, according to the calculations, to radiate huge amounts of energy at the shortest wavelengths. Since violet light has the shortest wavelength of any light in the visible spectrum, and radiation with an even shorter wavelength is known as ultraviolet, this came to be known as the "ultraviolet catastrophe." Something had to be wrong with the ideas behind those calculations. But what? The answer came from an unexpected quarter.

Enter the quantum

Max Planck was among the first generation of physicists to grow up with the statistical interpretation of thermodynamics developed by Maxwell and Boltzmann. He didn't like the idea at first; but at least he was familiar with Boltzmann's work when the opportunity came to apply it in a new and unexpected context.

Planck entered the University of Munich in 1874, and received his PhD, for a thesis on the second law of thermodynamics, in 1879. After a spell as a *Privatdozent* there, he moved to Kiel in 1885 and on to Berlin in 1888, becoming a full professor there in 1892 and staying in the post until he retired in 1926, to be succeeded by—Erwin Schrödinger. Planck did solid but unspectacular work, mostly in thermodynamics, and gave beautifully clear, precise lectures that were almost always standing room only events. He turned his attention to the puzzle of black body radiation in 1894—not, initially, in the abstract quest for understanding, but because a consortium of electricity companies had commissioned him to find out how to produce light bulbs that would provide the maximum amount of light from the minimum amount of energy. But once he got his teeth into the problem of black body radiation he became hooked, and worried away at it for years until he found an answer, publishing several important papers on the relationship between thermodynamics and electrodynamics along the way.

It was, indeed, Planck's grasp of statistical thermodynamics that led him to the resolution of the ultraviolet catastrophe—although he was not motivated to do so by the catastrophe as such. What Planck wanted to achieve was an understanding of the physical processes responsible for the exact shape of the spectrum of cavity radiation—the black body curve. He had two clues to help him. In 1896, Wilhelm Wien (1864–1928), working in Berlin, put forward an empirical rule—an equation derived more or less by trial and error—which gave an accurate representation of the short-wavelength side of the black body curve, and specified the wavelength of the peak for a given temperature. What became known as Wien's Law says that the temperature of a black body (in Kelvin, the absolute units of temperature) is given simply by dividing the number 2,900 by the wavelength of the peak (in micrometres). So if the peak is at a wavelength of 5 micrometres (0.005mm), the temperature of the object is 580K (307°C). But there was no explanation for why Wien's Law worked, and it only worked for the short-wavelength side of the peak—the equation was hopelessly inaccurate at longer wavelengths. Intriguingly, though, there was another equation which worked for the long-wavelength side of the peak, but was hopelessly inaccurate at shorter wavelengths, where it "predicted" the non-existent ultraviolet catastrophe. This second equation was the fruit of the work of two men, the British physicists Lord Rayleigh (1842–1919) and James Jeans (1877–1946), and became known as the Rayleigh–Jeans Law. Rayleigh came up with the original equation, which was later refined by Jeans, at the end of the nineteenth century; it essentially treats light as a classical wave, with all that that implies.

Planck's achievement was to find a single law, based on sound physical principles combined with statistical techniques familiar from thermodynamics, to produce a single rule which not only bridged the gap between Wien's Law and the Rayleigh–Jeans Law but explained the whole black body curve. But, as I have mentioned, the achievement came at a price.

Planck's starting point was to assume that the radiation from a black body is being produced by an array of "electromagnetic oscillators." He was careful not to specify the nature of these oscillators. By the 1890s it was well known that moving electric charges could produce electromagnetic radiation, and that light was a form of electromagnetic radiation; so it was natural to assume that electric charges oscillating to and fro produced the radiation emitted by black bodies. In the summer of 1900, Planck found that with a little mathematical juggling he could come up with an equation that smoothly joined the good bit of the curve described by Wien's Law to the good bit of the curve described by the Rayleigh–Jeans Law, providing a single equation to describe the entire black body curve; but at that stage he still had no physical basis for the equation. Then, having exhausted all other attempts at finding such a basis, in October that year he realized that what he needed was Boltzmann's statistical interpretation applied not just to the electromagnetic oscillators, but to energy—to the electromagnetic radiation itself.

Boltzmann's statistical approach to thermodynamics involves cutting energy up into little pieces mathematically (using the techniques of calculus), manipulating the pieces as required by the laws of statistics, and then adding the modified pieces together (integration) at a later stage of the calculation. Planck had never been a fan of this approach, but almost in desperation he tried it anyway, treating the electromagnetic radiation as made up of little pieces of energy instead of a smooth wave. The astounding result was that the equation he had already discovered empirically, which

quantum carries a lot of energy (relatively speaking), but it is hard to make such quanta and only a few of the atoms have enough energy to do the job. So although each quantum carries a lot of energy there are few quanta, and once again the overall energy radiated is low. It is only in the middle ground, corresponding to the peak in the black body curve, that it is possible to make a lot of quanta, each with a moderate amount of energy, so that the overall amount of energy radiated is high. And, naturally, for hotter objects there is more energy available, so it is easier for the atoms to make higher-energy quanta, and the peak shifts towards higher frequencies—in terms of colour, from red through orange towards blue.

In spite of this success, Planck's announcement did not change physics overnight. Nobody knew quite what to make of it, and to many people, at first, it did not seem to do much more than tidy up the work of Wien and Rayleigh (Jeans's refinement came a little later). Planck himself didn't quite know what to make of it either; many years later he wrote: "I can characterise the whole procedure as an act of despair . . . a theoretical interpretation had to be found at any price."[1] But the price Planck was willing to pay stopped short of accepting the reality of his energy elements as physical entities. He believed that what the equations were telling him was that the packages of electromagnetic radiation could only be emitted or absorbed in amounts of $h\nu$, but that the radiation itself was a classical wave. A rough analogy might be with the relationship between a cash machine and money. The cash machine can only dispense money in multiples of £10—what you might call "money elements." But other amounts of money, such as £55.76, exist in the world outside the cash machine.

Planck was forty-two when he made his unexpected discovery, and too set in his ways to make the next leap. It took a much younger man, with a fresh approach to physics, to look at the implications of accepting that energy elements—what we now call quanta—are real, physical entities; at the possibility that, as it were, we live in a world where money only exists in units of £10, inside or outside the cash machine. That younger man was Albert Einstein, who made his breakthrough in 1905, when he was twenty-six.

The quantum becomes real

Einstein graduated from the Swiss technical university, the ETH, in 1900, the year of Planck's breakthrough. But he had not covered himself in glory as a student, partly because he didn't bother to attend lectures and studied more or less what he liked. Graduating bottom of his class as a result, he was unable to pursue his ambition of working at the ETH for a PhD and becoming a *Privatdozent*, but eventually managed to get the now-famous job as a Technical Expert Third Class at the patent office in Bern. It was while working there that in 1905, a year before Schrödinger entered the University of Vienna, he produced an astonishing set of scientific papers, including a PhD thesis, the special theory of relativity, and one that he described that year in a letter to his friend Conrad Habicht as "very revolutionary"—a term he did not apply even to the special theory. It dealt with the relationship between radiation and atoms, and it proved that light quanta—photons—are real.

Einstein's jumping-off point was the work of Philipp Lenard (1862–1947), a German physicist who had carried out a series of experiments, starting in 1899, to investigate the way ultraviolet light shining on the surface of a metal in a vacuum could make the metal emit what were then still called cathode rays, but which are now known as electrons.

Lenard had discovered that the energy of the individual electrons produced by this so-called photoelectric effect does not depend on the brightness of the light shining on the surface. But it does depend on the wavelength, or frequency, of the light. This was weird. A faint light has less energy than a bright light, so surely a fainter beam of light should produce electrons with less energy. And what had the frequency of the light got to do with it? Einstein had the answer—or rather, he realized that Planck had given him the answer. If electromagnetic radiation with a particular frequency ν really is made up of a stream of particles each with energy $h\nu$, then every time one of these particles (quanta, now known as photons) knocks an electron out of the metal it does so with the same amount of energy. Turning the brightness of the radiation down reduces the overall energy of the beam, but only because it reduces the number of photons. For a particular frequency, each photon still has the same energy. So although fewer electrons are knocked out of the metal, each one still gets a kick with the same amount of energy, $h\nu$. As Einstein put it, "the simplest conception is that a light quantum transfers its entire energy to a single electron." The only way to change the amount of energy carried by each photon is to change the frequency of the light, not its intensity. Einstein again: "According to the assumption considered here, in the propagation of a light ray from a point source, the energy is not distributed continuously over ever-increasing volumes of space, but consists of a finite number of energy quanta localised at points of space that move without dividing and can be absorbed or generated only as complete units." In other words, if you had infinitely sensitive eyes and looked at a source of light from very far away, you would not see a faint, continuous overall glow, but individual flashes of light, with complete darkness in between, as individual quanta arrived at your eyes.

A century after Einstein made this claim, this is exactly what astronomers do "see," using sensitive electronic detectors to study the light from distant objects. They literally observe the arrival of individual photons, one after another. But in 1905, there were no experiments sensitive enough to support this claim. Indeed, even Lenard's experiments, although suggestive, were not really accurate enough to justify Einstein's far-reaching interpretation of the data, and it was widely felt that he had gone too far in drawing such a conclusion from imperfect evidence—not least since there was still the compelling evidence, from Young's experiment and others, that light travels as a wave. How could it be both wave and particle?

One person was so infuriated by what he saw as Einstein's nonsensical idea that he set out to prove him wrong. Robert Millikan was an American experimenter working at the University of Chicago. He was superbly skilful and among other things made the first accurate measurement of the charge on an electron. He was also, like all good scientists, willing to admit when he was wrong, and after an investigation lasting several years he concluded that the relationship between the energy of an ejected electron and the frequency of the radiation involved exactly matched Einstein's prediction. But he had no idea why. In 1916 he wrote: "The Einstein equation accurately represents the energy of electron emission under irradiation with light [but] the physical theory upon which the equation is based [is] totally unreasonable." And again in 1949, in an article in the *Reviews of Modern Physics*, he wrote: "I spent ten years of my life testing that 1905 equation of Einstein's and contrary to all my expectations, I was compelled to assert its unambiguous verification in spite of its unreasonableness."

Nevertheless, the experiments provoked by Einstein's "unreasonable" theory provided Millikan with an accurate measurement of Planck's Constant, demonstrating that it had real significance. After all, if you can measure it, it must be real. It is no coincidence that it was after Millikan reported his results that the Nobel Committee gave the physics prize for 1918 to Planck, and the prize for 1921 (the year Schrödinger took up his appointment in Zürich) to Einstein; Millikan received his prize in 1923. By then, the importance of quanta and quantum physics was no longer in doubt, and the understanding of the relationship between electrons, light, and atoms had been transformed by the application of quantum ideas. The man chiefly responsible for the breakthrough was the Dane Niels Bohr (1885–1962) whose own Nobel Prize, awarded in 1922, slotted in between those of Einstein and Millikan.

Inside the atom

Even while some physicists were still questioning the reality of atoms, others had been starting the process of breaking atoms apart to discover even smaller components of the material world. The process really began in 1896, with the discovery of radioactivity by Henri Becquerel (1852–1908), working in Paris. He found that some previously unknown emission from uranium could produce fogging of a photographic plate even though the plate was wrapped in a double layer of black paper to prevent light getting to it. This spontaneous emission came to be known as radioactivity, and other radioactive substances were soon discovered. Just a year after Becquerel's breakthrough, J. J. Thomson announced, in a lecture at the Royal Institution in London, the discovery that the radiation from a wire that is carrying an electric current in a vacuum tube is made up of a stream of electrically charged particles—what we now call electrons. The experimental study of radioactivity was swiftly carried forward by the Curies, Marie (1867–1934) and Pierre (1859–1906), at the Sorbonne; but the person who first appreciated what radioactivity involved, and then used radioactivity to probe the structure of atoms, was Ernest Rutherford (1871–1937), a New Zealander who worked in Canada and England.

Rutherford arrived in England in 1895 and worked for a time under Thomson at the Cavendish Laboratory in Cambridge. Under Thomson's influence, he became interested in atomic physics, and soon discovered that there are two kinds of radioactivity, one producing positively charged particles which he dubbed alpha radiation, and the other producing negatively charged particles which he called beta radiation. It soon became clear that beta particles are in fact fast-moving electrons, but the name stuck. When Rutherford identified a third kind of radiation, in 1900, it seemed natural to call it gamma radiation; gamma rays carry no charge and are a form of electromagnetic radiation—essentially, very energetic X-rays.

By the time he discovered gamma radiation, Rutherford had moved on to McGill University in Montreal, where he worked with the English chemist Frederick Soddy (1877–1956). Together, they discovered that the process of radioactivity involves the transformation of atoms of one element into atoms of another element, and that this process (known as decay) takes place on a characteristic timescale, different for each radioactive substance, measured by what is called its half-life. For any quantity of a particular radioactive element, in one half-life half of the sample (half the atoms) decays into another element by emitting alpha or beta radiation. For example, radium, one of the radioactive elements discovered by the Curies, decays with a half-life of 1,600 years, by emitting an alpha particle from each atom that decays, to produce a gas called radon. Alpha particles themselves turned out to be identical to atoms of helium, the second-lightest element, from which two electrons have been removed, leaving them with two units of positive charge. Each alpha particle has more than seven thousand times as much mass as an electron, roughly the same as the mass of four hydrogen atoms. But even before he knew what an alpha particle was, or how such particles could be ejected at high speed from radioactive atoms such as those of uranium or radium, Rutherford was able to use them to study the structure of atoms.

Back in England, in 1907 he became Professor of Physics at the University of Manchester, and in 1908 received the Nobel Prize for his work on the transmutation of the elements. The Nobel Committee regarded this as a branch of chemistry, so he was awarded the chemistry prize, even though Rutherford regarded himself as a physicist and once famously said: "All of science is either physics or stamp collecting." The year after he received the Nobel Prize,

Rutherford suggested an experiment, actually carried out by two of his junior colleagues in Manchester, Hans Geiger (1882–1945) and Ernest Marsden (1889–1970), that provided the first insight into the structure of the atom.

Geiger and Marsden used alpha particles from a radioactive source to bombard thin sheets of metal foil, and a detector devised by Geiger and Rutherford (the precursor of the famous Geiger counter) to monitor where the alpha particles went after they hit the metal. To their astonishment, they found that although most of the particles went right through the foil as if it were not there, some were deflected at large angles and occasionally one bounced right back from it, like a ball being hit against a brick wall. Rutherford explained this by suggesting that an atom consists of a tiny central kernel (soon to be dubbed the nucleus), which is positively charged (so it repels positively charged alpha particles) and contains most of the mass of the atom, surrounded by a cloud of electrons, which take up most of the space but have very little mass. An alpha particle can brush past the electrons with hardly any effect, but if it just happens to hit, or pass close by, a nucleus it will be deflected violently. By careful study of the statistics of which proportions of the alpha particles were deflected at different angles, Rutherford was even able to calculate the relative size of the nucleus. An atom is typically about 10^{-8} centimetres (one hundred-millionth of a centimetre) across, and the nucleus is typically about 10^{-13} cm across (one hundred-thousandth the diameter of the atom).

Rutherford announced this nuclear model of the atom in 1911, although he only introduced the term "nucleus" in this context a year later. It neatly explained the way alpha particles (now regarded as the nuclei of helium atoms) are deflected by atoms. But it posed one big problem—since electrons have negative charge, and the nucleus of an atom has positive charge, and opposite charges attract, why don't all the electrons fall into the nucleus? By great good fortune, someone who was able to see the answer to this puzzle just happened to be visiting Manchester from Denmark for a few months, at exactly the time Rutherford's team was making this breakthrough. Niels Bohr was to become one of the most influential of the founding fathers of quantum theory.

Tripping the light fantastic

There was another puzzle about the relationship between light and matter, which dated back to the early nineteenth century, when Thomas Young was promoting the wave model of light. One of Young's friends, William Wollaston (1766–1828), having made a fortune by inventing a technique for producing pure platinum, was able to indulge his passion for science as an independent investigator. He was a strong supporter of the wave model of light, and was the first person to notice that the light from the Sun, when spread out into a spectrum using a prism and studied through a microscope, is marked by a multitude of dark lines. These lines were studied in more detail by the German physicist Josef von Fraunhofer (1787–1826), who developed the "prism spectrometer" into a refined scientific instrument, later known as the spectroscope. He used this instrument to study the spectrum of the Sun in the second decade of the nineteenth century; these dark lines within the spectrum are now known as Fraunhofer lines. But what causes them?

In the decades that followed, researchers including Fraunhofer himself, Robert Bunsen (1811–99), and Robert

Kirchhoff, established from experiments in their laboratories that each element produces its own set of lines in the spectrum. Each line corresponds to a particular frequency, or colour, of light, and each element radiates at its characteristic frequencies when hot, but absorbs light at those frequencies when cold. The pattern of lines associated with each element is unique, and as distinctive as a fingerprint or a bar code. For example, when sodium is hot it radiates at particular frequencies in the yellow-orange part of the spectrum. Some street lights contain a gas which includes sodium compounds, which is why their light is orange. But if light passes through a cool gas containing the same sodium compounds, the sodium absorbs light in the yellow-orange part of the spectrum.

Spectroscopy—the analysis of these patterns—turned out to be a highly useful tool for chemists in analysing the composition of different substances, even though nobody knew how the lines were produced. And it meant that there must be a multitude of different elements in the relatively cool atmosphere of the Sun, absorbing light passing through it from the much hotter surface below, each set of black lines corresponding to an element. The crowning triumph of this new branch of science came when the British astronomer Joseph Lockyer (1836–1920) used the discovery of a set of lines in the solar spectrum that did not correspond to any known element to infer that they must be associated with a "new" element, which he called helium; it was only in 1895, nearly thirty years after Lockyer reached this conclusion, that helium was identified on Earth.

Bohr's genius was to solve the mystery of spectroscopy, locate the origin of Planck's (or Einstein's) quanta, and explain why atoms don't collapse—all in one neat package. Physics would never be the same again.

Bohr began to develop what became known as the Bohr model of the atom during his first visit to Manchester (he came back for a longer visit between 1914 and 1916), but completed it back in Copenhagen in 1912. The idea was published a year later, around the time Schrödinger was seriously thinking of giving up physics in order to marry Felicie Krauss. It contained the essence of Bohr's approach to any problem—a willingness to combine different ideas from different areas of physics in a patchwork to make a working model, even if the different patches didn't, at first sight, seem to belong together. Once this gave him a rough idea of a model, he (or someone else) could adjust the pieces to make a snugger fit and (hopefully) come up with something better.

The first piece of Bohr's patchwork came from classical physics. It "explained" Rutherford's discoveries about the structure of the atom by postulating that electrons are "in orbit" around the central nucleus, in a way reminiscent of the way planets orbit the Sun. According to classical physics (Maxwell's equations), electrically charged particles moving in such orbits ought to radiate electromagnetic energy continuously, and spiral down into the nucleus as a result. Bohr ignored this dogma, and for his next piece of the patchwork turned instead to quantum physics. He suggested that the electrons could only emit or absorb energy in definite lumps—Planck's quanta. So they could not spiral steadily inward. Instead, they could only jump, outward or inward, by definite amounts, corresponding to $h\nu$. Only certain orbits, corresponding to certain amounts of energy, were allowed, and electrons could move between those orbits, jumping inward if they released a quantum of energy and outward if they absorbed a quantum of energy. (The rules about which energy levels were allowed were initially worked out empirically, and labelled by a set of numbers simply known as "quantum numbers.") It would be as if the Earth suddenly jumped either inward to the orbit of Venus, or outward to the orbit of Mars, without—and this is a crucial point that would later stick in Schrödinger's throat—passing through the space in between. But still, why didn't all the electrons in an atom jump down into the lowest orbit? Because, said Bohr, plucking another rabbit out of the hat, the inner orbits were already full! There must be a limit to how many electrons could occupy each orbit, and once that limit was reached any more electrons had to occupy orbits farther out from the nucleus.

The idea was so outrageous that it might have been laughed out of court. But Bohr had an ace up his sleeve—or rather, a pair of aces. The first is only tangentially relevant to the story of Schrödinger as a reluctant quantum pioneer, so I will not go into details here (I have described it fully in my book *In Search of Schrödinger's Cat*). Briefly, it is that Bohr's model of the atom gave the first plausible explanation of why atoms can only join together to make molecules in certain ways, by forming so-called "bonds"—why, for example, a water molecule always consists of a single oxygen atom joined to two hydrogen atoms, and we never find a molecule composed of a single hydrogen atom joined to two oxygen atoms. It made chemistry a branch of physics. The fact that Bohr's model was later superseded by a better model of chemistry doesn't matter any more than the fact that Newton's theory of gravity was superseded by Einstein's general theory of relativity.

The second ace, however, is directly relevant to my story. When the appropriate numbers were put in, in particular from Planck's radiation equation, the Bohr model predicted the lines in the spectrum, produced by electrons jumping between energy levels, known to correspond to the simplest element, hydrogen (the calculations were too complicated to work out theoretical spectra for heavier elements). It also predicted some lines that were not seen, but that was a relatively minor problem that could be (and was) tidied up later. For all its patchwork nature, Bohr's model of the atom was the first to explain spectral lines, as well as the first to explain the observations made by Rutherford's team. Even though progress in theoretical physics was slowed by the First World War, Bohr's ideas were refined and improved by others, not least by Albert Einstein once he had completed his masterwork, the general theory of relativity, in 1916.

Einstein again

In its original form, like Planck's equation for black body radiation, Bohr's atomic model did not say whether electromagnetic radiation could only exist in the form of discrete chunks (quanta), or whether it could only be emitted or absorbed in discrete chunks, like money from the cash machine. And what, exactly, was the relationship between Bohr's model and Planck's equation? It was Einstein who provided the answers—and more.

The calculations for the hydrogen atom had been simple enough to work with because each hydrogen atom has a single positively charged particle (a proton) in its nucleus, and a single negatively charged particle (an electron) "in orbit" around it, jumping between the permitted energy levels in a way crudely analogous to a ball bouncing up and down a flight of steps. Einstein's contribution was to find a way to describe mathematically what is going on when many electrons are jumping around the energy levels in large numbers of much more complicated atoms, as if a box of balls had been emptied on to the staircase and they were all bouncing around at once. Naturally, as the leading proponent of the idea that photons are real, he argued that when an electron jumps down an energy level it emits a photon with the same energy as the difference between the two levels, which in turn corresponds to a precise frequency of light (given by Planck's formula), and when an atom absorbs a photon one of its electrons jumps up an energy level; but this is only possible if the atom possesses two energy levels separated by a gap just the right size to absorb radiation with that frequency. Einstein's overall description of what was going on was based on statistics. He used standard statistical techniques to work out the probability that an atom with a particular set of quantum numbers would "decay" into a state with lower energy and a correspondingly different set of quantum numbers, emitting a photon in the process. Adding up the contributions of these photon emissions, he was able to calculate the overall radiation from a large number of atoms in a hot object. The result was Planck's equation for the black body curve.

It was this discovery that established the acceptance of Bohr's model of the atom, and finally completed the first quantum revolution. But it contained a ticking time bomb. The whole edifice rested upon probabilities. Einstein had introduced into quantum physics the idea that you could never calculate with certainty just what a particular atom (or any other quantum object) would do next. You might say that there was a 1 in 3 chance, for example, that a particular kind of atom with a particular set of quantum numbers would emit a photon with a particular frequency in the next ten minutes. But it might not decay for days, or years. The calculations worked because large numbers of atoms were involved. If you had a million atoms like this you could be sure that almost exactly 333,333 of them would decay in the next ten minutes—but you could never tell in advance which atoms would decay and which ones would not. In the 1920s, probability and uncertainty became integral parts of the second quantum revolution, leading Schrödinger and (ironically) Einstein to have serious doubts about the whole enterprise. Even when he introduced the idea in his paper describing his breakthrough in understanding the quantum theory of radiation, Einstein had described this as "a weakness of the theory," expressing concern that "it leaves time and direction of elementary processes to chance." But the old quantum physics, as it came to be known, still had a couple of contributions to make before it was blown away. And Einstein himself was still on the case.

One of the other ideas in his quantum theory of radiation paper, published in 1917, was a way in which atoms might be encouraged to emit radiation. If many atoms, perhaps in a crystal, were prepared in a high-energy state—in effect, with an electron in each atom sitting on a particular high step of the staircase—it ought to be possible to trigger them into decaying into the lower-energy state by giving them a nudge. And the most efficient way to provide such a stimulus would be to nudge the atom with a photon that had exactly the same energy as the jump you wanted to trigger, causing a kind of resonance. If enough atoms were available to be jostled in this way, a single photon could trigger one of them to decay without being altered itself, so that there would now be two of them to trigger the next pair of atoms, then four, eight, sixteen, and so on in an exponentially growing cascade. The result would be an intense beam of pure radiation, all with exactly the same frequency (colour), amplified up from the original stimulus. Decades later, this idea of "light amplification by stimulated emission of radiation" became a practicality, and gave us the term "laser." Lasers are pure quantum physics put to practical use.

But there was still more in Einstein's paper. Any moving particle in the everyday world carries both energy and momentum—momentum is a measure of how much impact a moving object has on anything with which it comes into contact, and depends on both the mass of the object and its velocity, so that a light object travelling very fast can hit with an impact as great as that of a heavy object moving more slowly. If photons were real particles that carried energy, they must have momentum. Einstein showed that, although photons are a special case because they always travel at the speed of light, usually denoted by the letter c, the mathematics of Planck's description of the black body curve and his own description of the interaction between atoms and radiation were only consistent with one another if each photon with energy E has a momentum given by dividing E by c.

This was a dramatic claim, and encouraged experimenters to test Einstein's prediction. The one who succeeded was the American physicist Arthur Compton (1892–1962), who measured the transfer of momentum from X-ray photons to electrons in so-called "scattering" experiments late in 1922 (the results were published in 1923). He received the Nobel

Prize for what became known as the Compton effect in 1927. It was well deserved, since the Compton effect provides definitive proof of the reality of photons and the accuracy of quantum physics (even the old quantum physics!) as a description of the subatomic world.

But buried within all this was a mystery that would be one of the triggers of the second quantum revolution, and would inspire Schrödinger's masterwork. Planck had shown that the energy of a photon is related to its frequency by the equation $E = h\nu$. Einstein had shown that the momentum of a photon, usually denoted by the letter p, is related to its energy by the equation $p = E/c$. In other words, $p = h\nu/c$. The momentum of a photon, which had previously been regarded as purely a particle property, is directly related to its frequency, which had previously been regarded as purely a wave property. What was going on? It was against this background that Schrödinger arrived in Zürich, and at first consolidated his position as a sound but unspectacular scientist.

CHAPTER FIVE

Solid Swiss Respectability

Schrödinger arrived in Zürich in the autumn of 1921, and soon settled into the rhythm of life as a respectable professor in a respectable university in a respectable country. It would be five years before he made the breakthrough that sealed the success of the second quantum revolution—years in which he gave no hint of what was to come.

Zürich was not quite a backwater of science in the 1920s, but it certainly was not a centre of excellence in research, in spite of having two respectable institutes of higher learning. The town's location on a lake and a river, where north–south and east–west trade routes crossed, meant that there had been a settlement there since pre-Roman times, and by the end of the first millennium A.D. it had become an important centre of trade, politics, and religion. In 1218 Zürich acquired the status of a free city, and in 1351 it joined the Swiss Confederation and was involved in the wars and tribulations that eventually (in 1648) led to Swiss independence from the Habsburg

Empire. A century earlier, Zürich had become a centre of the Protestant Reformation and a safe haven for Protestants fleeing persecution in other countries, boosting its cultural and intellectual diversity and making the city one of the intellectual centres of the German-speaking world in the eighteenth century. The modern Swiss state dates from the time of the French invasion of 1798, its constitution evolving through various modifications in the nineteenth century.

The university and the ETH

The University of Zürich was established in 1833, but although the teaching of physics got started on a good footing, very little research in the subject was carried out there until the arrival of Rudolf Clausius (1822–88) as professor in 1857. He didn't have to come far to take up the position, since he had been a professor at the new Zürich polytechnical institute, or Polytechnikum, since 1855—indeed, he retained that post alongside his role at the university. This was typical of the close links between the two institutions.

The Polytechnikum had a curious background. The original plan, dating back to the 1790s, had been to establish a technical institute along the French lines, like the École Polytechnique in Paris; but by the time the scheme came to fruition, in 1855, the model that emerged was more like that of a German *technische Hochschule*, reflected in its full name, the Eidgenössiche Technische Hochschule, or Swiss Federal Institute of Technology. The presence of both ETH and university in the same city is a result of the structure of the Swiss Federation. The ETH is a federal institution, and its natural home is the largest city in Switzerland, which happens to be Zürich; but the university is a cantonal institution governed by the local community. So the ETH is—or certainly was at that time—the more important of the two, even though to modern ears the name "university" has more prestigious overtones. For the next fifty years, the ETH, rather than the university, was the leading centre of physics in Zürich, and the university's only period of excellence in the field came during the tenure of Clausius, from 1857 to 1867, when he held the ETH post as well.

Clausius had studied and worked in Berlin before moving to the ETH. His great claim to fame is his recognition of the fundamental law of nature that heat cannot flow of its own volition from a colder object to a hotter object; or, as he put it in a paper published in 1850, "it is impossible by a cyclic process to transfer heat from a colder to a warmer reservoir without net changes in other bodies." This is the simplest formulation of the second law of thermodynamics. And it was Clausius who, in 1865 while he was working in Zürich, introduced into physics the term and mathematical concept of "entropy" as a precise measure of disorder. But in 1867 Clausius moved on, first to Würzburg and then to Bonn, and physics in Zürich languished.

But mathematics flourished, at least at the ETH, where its most famous pupil, Albert Einstein, arrived in 1896 to take the course intended to produce high-school teachers of science. Among his own teachers was Hermann Minkowski (1864–1909), who famously once described young Albert as "a lazy dog who never bothered about mathematics at all," but soon came to appreciate the achievements of his former student and in 1908 came up with the brilliant idea of explaining Einstein's special theory of relativity, published just three years earlier, in terms of the geometry of four-dimensional "spacetime." Minkowski's geometrization of the special theory was a major factor in its rapid acceptance, and would later become a powerful tool in understanding the general theory of relativity. Just a year after Minkowski presented the idea, Einstein received his first academic appointment, at the University of Zürich. But the resulting second flowering of physics at the university was to be even more short-lived than the first.

Einstein's growing reputation brought a stream of eminent visitors to Zürich to discuss physics with him, and one of his research interests at the time, applying quantum theory to the understanding of specific heats, led to his being invited to present a paper on the subject to a conference in Brussels in the autumn of 1911. By the time he presented that paper, however, Einstein had moved on from Zürich to Prague, and his successor, Peter Debye (1884–1966), had recently been installed. Debye was an excellent physicist who would carry out important work on atomic and quantum theory, but he lasted only a year in Zürich before moving on to Utrecht and being replaced by Max von Laue,[1] who was by then a well-established scientist; his idea of investigating crystal structure using X-rays had been suggested in the spring of 1912 and confirmed by experiments later the same year. But von Laue was too big a fish for Zürich to hold on to, and he moved to Frankfurt-am-Main in the summer of 1914, a few months before he received the Nobel Prize.

Meanwhile, in 1912 Einstein had returned to Zürich to take up a special professorship at the ETH, free from any general lecturing duties, although he was required to give lectures and seminars for advanced students. As well as other work, including research in statistical mechanics, during his time as a professor at the ETH Einstein was deeply involved in developing his theory of gravity—what would become the general theory of relativity. But he too was lured away in 1914, this time to a prestigious and well-paid post in Berlin, where, free from any teaching duties at all, he was able to complete this work over the next two years.

All this left Zürich distinctly light in theoretical physics expertise when the First World War began, and, partly because of the restrictions on movement between countries caused by the war, even though Switzerland was neutral, the situation did not improve for the rest of the decade. As far as teaching was concerned, two young *Privatdozenten*, Simon Ratnowsky (1884–1945) at the university and Mieczyslaw Wolfke (1883–1947) at the ETH, did a sterling job, updating the lectures to include the latest advances in quantum theory. But their research was solid second-division stuff, not material likely to attract the attention of the Nobel Committee. Towards the end of the decade they began to be overshadowed by the work of their colleagues, notably Paul Epstein (1883–1966), a quantum theorist who arrived at the university from Munich in 1919, and Hermann Weyl (1885–1955), who had been at the ETH since 1913 and whose growing reputation had reached a peak with a masterful book on relativity theory, *Space–Time–Matter*, published in 1918. In 1920 Peter Debye returned to Zürich as Director of the Physics Institute at the ETH, and when Wolfke went back to his native Poland the following year to take up a post in Warsaw, the way seemed clear for either Ratnowsky or Epstein to be promoted to the chair of theoretical physics at the university, which had been vacant since von Laue's departure nearly a decade earlier.

But to the surprise of many at both Zürich institutions, it was a relatively unknown outsider, Erwin Schrödinger, who got the job.

In fact, the appointment made sense. As I have mentioned in an earlier chapter, the commission in charge of making the appointment had tried to get von Laue back, but were unable to meet his financial expectations, and Epstein ruled himself out in March 1921 by accepting the post in Leiden. The physics faculty then recommended Schrödinger on the grounds of his excellence as a teacher and his versatility, covering "the fields of mechanics, optics, capillarity, electrical conductivity, magnetism, radioactivity, gravitation theory, and acoustics"—a list in which quantum theory is conspicuous to modern eyes by its absence. One reference also mentioned that "he has a nice wife." When Schrödinger arrived in Zürich, though, they must have wondered if he was up to the job.

Personal problems and scientific progress

The cause for concern was Schrödinger's health. Mentally and physically exhausted by the events of the years since 1918, which included the death of both parents and a grandfather, as well as the economic difficulties and the problem of simply getting enough to eat, he was forced to abandon his lectures less than halfway through his first term in Zürich by a severe bout of bronchitis. Respiratory problems persisted through the winter, culminating in the diagnosis of a mild case of tuberculosis.

In the 1920s, the only treatment for TB was to rest, preferably at high altitude, and hope for the best. The idea behind this "cure" was that high altitude encourages the body to produce red blood cells, which were thought to fight the infection; in fact, in so far as the cure did any good at all, it was probably because, as we now know, the TB bacteria need plenty of oxygen, which is in short supply at high altitude, in order to thrive. Whatever the reason, Schrödinger's stay at the Alpine resort of Arosa, near Davos, proved effective. He was there—with Anny and a splendid cook who prepared all his favourite meals—for nine months, returning to the university only in November 1922, well after the start of the academic year. Suntanned but weak and liable to tire easily, he was ready to pick up his teaching duties, although lacking any drive for his own research. Over the next few years, Schrödinger returned to Arosa several times, including during the summer of 1925 and the following Christmas, for holidays and for the benefit of his health.

Remarkably, though, Schrödinger had completed two scientific papers while convalescing in Arosa in 1922. One was a mundane piece of work on specific heats. But the other contained a nugget that would not be appreciated for the pure gold it was until 1926, after Schrödinger made the breakthrough for which he received the Nobel Prize.

The problem Schrödinger addressed in this second paper of 1922 was the way in which the orbits allowed for an electron in the Bohr model of the atom are quantized. He started from an approach developed by Weyl in his book, which had already become a standard text, and found that for the *n*th orbit out from the nucleus a property known as the "unit of measure" peaks *n* times around the orbit, with troughs in between. As Schrödinger put it: "If the electron in its orbit carried along with it a 'distance,' which is transferred unchanged during the motion, then the measure of this distance would—if one started from an arbitrary point on the orbit—always be multiplied by an integral multiple of ev, exp(h/γ) whenever the electron returns approximately to its initial position."

Schrödinger did not attempt to find a physical explanation for this mathematical result, although he did say that it was "hard to believe" that there was no "deeper physical meaning" to it. But this is, of course, exactly like the picture of a standing wave surrounding the nucleus. If Schrödinger had not been so debilitated by his illness and exhausted by his teaching, he might well have come up with his Big Idea by the end of 1922, instead of in 1926.

There was also a more subtle, but no less important, idea tucked away at the end of the paper. Schrödinger pointed out that the equation he had found contained a number which could have either of two values. One of these solutions was an ordinary number—what mathematicians call a "real" number. But the other was what they call an "imaginary" number, meaning that it is a real number multiplied by the square root of −1, which is denoted by the letter i. Numbers such as 1, 2, 3 . . . are real; numbers such as i, $2i$, $3i$ . . . are imaginary. Such "imaginary numbers" can be manipulated mathematically in just the same way as real numbers, and sometimes crop up in situations where they are regarded as physically meaningless and ignored. One place where imaginary numbers cannot be ignored, though, is in the equations that describe the behaviour of waves. But what we see clearly with hindsight was no more than a vague hint of what was to come to even the sharpest minds of 1922.

Schrödinger's teaching load was enough to slow the honing of his own sharp mind as he gradually recovered from his illness—in a reversal of the situation in most universities today, in the 1920s the senior professors did most of the teaching, and in Schrödinger's case this amounted to eleven hours per week. He was also distracted by the need to prepare and present his formal inaugural lecture as a professor, more than a year after his arrival at the university. The subject of the talk, given on 9 December 1922, was one close to his heart, if not quite what the audience might have anticipated.

Physics and philosophy

The title of the inaugural lecture was "What Is a Physical Law?" Schrödinger took as his starting point Boltzmann's ideas about thermodynamics, according to which the second law is merely a statistical rule that applies to very large numbers of atoms and molecules, but individual atoms and molecules are unaware of it (so that there is no "arrow of time" for a collision between a single pair of atoms). Schrödinger was influenced by Franz Exner, with whom he had worked before the war, who had taken the extreme position of arguing that all natural events are an accumulation of chance occurrences, and that at the level of individual atoms there are no "laws of nature."

Schrödinger's own position was also influenced by Einstein's discovery that light carries momentum. If an atom ejected a particle which carried momentum, the atom ought to experience a recoil, just as a gun "kicks" backward when a bullet is fired from it. This obeys a rule known as the law of conservation of momentum, verified in countless experiments, which says that the momentum carried by the bullet in one direction is exactly equal to the momentum carried in the opposite direction by the recoiling gun. A small bullet moving at high speed has the same amount of momentum as a large gun recoiling more slowly. But if an atom radiates a spherical wave of light equally in all directions, as physicists thought it did in 1922, how can there be any recoil?

In a letter to Weyl written while he was preparing his inaugural lecture, Schrödinger said that he was driven to conclude that at the level of individual atoms the law of conservation of momentum did not hold. And if that law did not hold, why should any of the other laws of physics? Schrödinger was doubting the very idea of causality, seemingly a bedrock of science. In his own words: "Physical research has clearly and definitely shown that chance is the common root of all rigid conformity to Law that has been observed, at least in the overwhelming majority of natural processes, the regularity and invariability of which have led to the establishment of the postulate of universal causality."[2]

This was completely counter to the view of most physicists, then and since, although Schrödinger's closest friend in Zürich, Hermann Weyl, shared some of his ideas. The mainstream remained fully convinced that the basic laws such as the conservation of momentum apply with full force to collisions between individual pairs of atoms or molecules (and, indeed, to subatomic particles). Even in 1922, Schrödinger was skating on thin ice with these ideas. It had been known for decades that the laws of conservation of energy and momentum, in particular, are intimately bound up with our understanding of the nature of space and time. Specifically, if the laws are invariant, in the sense that they apply anywhere in space and at any moment in time, they have to be absolute truths, not statistical half-truths. And Schrödinger knew this! In the same lecture in which he suggested that physical laws might be only statistical in nature, he admitted that "the

Einstein theory [that is, the special theory of relativity] in no uncertain terms makes plain the absolute validity of the energy-momentum principles." For once, in that lecture Schrödinger the philosopher clouded the judgement of Schrödinger the physicist. But within a few years he had completely reversed his position, and after the second quantum revolution he became, along with Einstein, one of the most outspoken opponents of the idea that chance played a part in determining the outcome of events at the level of atoms and subatomic particles.

With the inaugural lecture out of the way, Schrödinger settled into a quiet but steady life as a university professor. He maintained and developed his interests in colour vision, culminating in the encyclopedia article completed in 1925 and published in 1926, and in statistical mechanics; and alongside this work he began to make minor, but significant, contributions to atomic and quantum theory, moving on from the paper completed during his "cure" at Arosa. His output increased as his health improved, although for a long time he was troubled by a persistent cough. Even though he published nothing at all in 1923, between 1922 and his breakthrough work in 1926 Schrödinger published six papers on statistical mechanics, five (including the encyclopedia article) on colour vision, four on specific heats, four on atomic structure, and one on relativity theory. It's a sign of Schrödinger's status in the physics community at this time that in 1924 he was invited to attend a prestigious scientific gathering in Brussels, the Solvay Congress, but not to present a paper there; he was highly regarded, but not one of the élite.

Schrödinger seems to have had a knack of believing two contradictory things at once (or at least, in turn), which was an invaluable ability for anyone trying to get to grips with the quantum world in the 1920s. Although in his inaugural lecture he firmly espoused the idea of light travelling as a wave, earlier in 1922 he had published a paper which tidied up a loose end in quantum physics (the term "quantum mechanics" was not introduced until 1924, in a paper by Max Born). This gave a full mathematical description of the Doppler effect (the way the wavelength of radiation is affected by motion) within the context of the special theory of relativity, but based entirely on the idea of light quanta (photons) carrying momentum. This was before Compton's experimental confirmation of the physical reality of Einstein's light quanta. But he still accepted the evidence from experiments such as Young's that light travels as a wave. In the letter to Peter Debye mentioned earlier, he referred to this as a "fatal dilemma" which led him to conclude that momentum is not conserved in atomic processes.

The idea that the laws of physics might only be statistical rules fitted in with Schrödinger's longstanding interest in statistical mechanics. This continued to be part of his research, as well as his teaching, in the first half of the 1920s. As well as his published work on specific heats, he planned, but never completed, a book on "molecular statistics." In 1924, when the Indian physicist Satyendra Nath Bose (1894–1974) came up with a new way of applying statistics to light quanta, and Einstein found other applications of the new idea, it was inevitable that Schrödinger would seize upon it and delve into the implications. That delving would lead him to his masterwork—driven by inspiration not just from physics, but also from his private life.

Life and love

Life in Zürich was pretty good. The Schrödingers moved in a circle of academics and intellectuals. There was a lively night life, which didn't particularly appeal to Erwin, but also theatre and the opera, which certainly did. When Anny wanted to go to one of the fine balls in the city, she went with one or another of Erwin's friends, while he stayed at home. In summer, groups of friends would take the steamer to the island of Ufenau for picnicking and swimming excursions, giving Erwin a chance to admire the pretty girls and engage in some mild flirtation. Something more than mild flirtation soon developed between Anny Schrödinger and Weyl. In an echo of *fin de siècle* Vienna, liaisons between married members of the academic set and others who were not their spouses were regarded as normal, and nothing to make much fuss about. In this particular circle, Anny Schrödinger and Hermann Weyl (known as Peter among his friends) soon became an item, as did Weyl's wife, Hella, and the physicist Paul Scherrer (1890–1969), one of Debye's protégés, who later became head of physics at the ETH. Erwin enjoyed several brief affairs, but the relationship between Anny and Weyl went deeper, and there was even talk of divorce, partly fuelled by Erwin's disappointment that his marriage with Anny was childless. But, as we shall see, they found other ways to resolve the situation.

Schrödinger also toyed with the idea of getting a "divorce" from the University of Zürich. The seed was sown when he attended a meeting of German scientists in Innsbruck, in September 1924. With Germany then at the forefront of research in physics, many of the big names involved in what was now coming to be known as quantum mechanics were there. Once again Schrödinger, although actively involved in the scientific discussions, did not make a formal contribution to the proceedings; but he was happy to be back in Austria, where, as in Germany, things seemed to be settling down at last after the turmoil of the First World War, in spite of some recent unrest—it was now five months since Adolf Hitler (a fellow Austrian) had been sent to gaol, where he was writing *Mein Kampf*, for his part in the Munich "beer-hall putsch." Oblivious to how this would affect his life, when Schrödinger was invited in the summer of 1925 to take up a professorship in Innsbruck he was very tempted, not least by what he referred to as the "gilded" memory of his visit the previous autumn. "The Swiss," he wrote to fellow physicist Arnold Sommerfeld (1868–1951), "are far too cheerless."

His correspondence reveals that in his own mind Schrödinger had decided by January 1926 to turn down the offer, but that he strung out the negotiations in order to obtain leverage for an improvement in the conditions in Zürich—a new blackboard and a larger budget for the physics library. He officially declined the invitation to Innsbruck only in March 1926, after the first of his ground-breaking papers on quantum mechanics had been published.

While this was going on, Schrödinger became involved in a controversy that could have had serious repercussions, and cast a permanent slight pall over his reputation. Einstein's special theory of relativity is based upon the postulate that the measured speed of light is the same for all observers, no matter where they are or how they are moving, and this is borne out by a variety of experiments. The key practical evidence, as of the 1920s, came from a series of experiments carried out by Albert Michelson (1852–1931) and Edward Morley (1838–1923), dating back to the 1880s. In 1921, a similar experiment was carried out at the top of Mount Wilson, the site of an astronomical observatory in California, and seemed to show a slightly different result from measurements made at sea level. It was this claim that prompted Einstein to express his disbelief by making his famous comment "The Good Lord is subtle, but he is not malicious."

But others were malicious. Anti-Semites in Germany had long rejected Einstein's "Jewish" theory, and they seized upon this "evidence" of his fallibility. One of the leading physicists in Germany, Philipp Lenard, was another Austrian who shared some of these unsavoury views. In 1922, when the German foreign minister, the Jewish Walther Rathenau, had been assassinated, Lenard, as Director of the Physics Institute in Heidelberg—which became notorious as a centre of right-wing activism—had refused to have the flag lowered to half-mast, and had to be taken into protective custody to save him from the resulting wrath of the mob. The extent of the undercurrent of unthinking anti-Semitism rife at the time can be gleaned from Schrödinger's own comment on the Mount Wilson experiment, that it "is very important, but it has been played down in Jewish circles of physicists." He urged that the experiment be repeated on the Jungfrau, and was quite unfazed when it was suggested that a member of the Heidelberg Institute would be the right man for the job.

Others doubted whether any of Lenard's protégés would give an honest appraisal; but Schrödinger furiously insisted that the scientific truth would out, whatever the politics of the experimenter. In the event, a Heidelberg team did do the experiment and it proved that Einstein was correct. So, in a sense, was Schrödinger; but although he seems to have acted from an apolitical standpoint, the incident established a perceived connection between him and the political right.

But at the end of 1925 Schrödinger's mind was focused on philosophy, rather than politics. He was now in his late thirties, far past the age when most great theorists make their contribution to science, and settled in what could be a job for life in a safe, secure, but dull setting. It was against this background that he sat down and wrote a summary of his philosophical views on the nature of the world, which was published much later as *Meine Weltansicht* (translated into English as *My View of the World*).

"My world view"

The opening part of Schrödinger's 1925 essay shows that he was well aware of what was happening in Europe at the time, and still deeply troubled by his own experiences. He wrote: "A sort of general atavism has set in; western man is in danger of relapsing to an earlier level of development which he has never properly overcome: crass, unfettered egoism is raising its grinning head, and its fist, drawing irresistible strength from primitive habits, is reaching for the abandoned helm of our ship."

It is hardly surprising that Schrödinger, appalled by this vision, should be drawn to the Vedantic view of the world. In *Meine Weltansicht*, he describes the idea of "a soul dwelling in the body as in a house, quitting it at death, and capable of existing without it" as "naively puerile," and asks four questions which, he says, cannot be answered "yes" or "no" but "lead one in an endless circle":

> Does there exist a Self?
> Does there exist a world outside Self?
> Does this Self cease with bodily death?
> Does the world cease with my bodily death?

The heart of the essay is Schrödinger's version of "The Vedantic Vision," which resolves these questions by arguing that there is only one consciousness, and that we (and, indeed, the rest of "nature") are part of it, like the different facets of a many-faceted jewel:

> It is not possible that this unity of knowledge, feeling and choice which you call your own should have sprung into being from nothingness at a given moment not so long ago; rather this knowledge, feeling and choice are essentially eternal and unchangeable and numerically one in all men, nay in all sensitive beings . . . you—and all other conscious beings as such—are all in all.

This one universal being is what is known as Brahman. "It is the vision of this truth," says Schrödinger, "which underlies all morally valuable activity."

In the rest of his essay, Schrödinger turns his attention to evolutionary biology, consciousness, and the process of heredity. Although he strikingly asserts that what his contemporaries delight in calling the age of technology "will in some later time" be described "as the age of the evolutionary idea," the only real interest in this part of his essay today is that it shows his fascination with the process of inheritance, which would reach fruition two decades later with his book *What Is Life?*. But the essay was not published until 1961,[3] when Schrödinger added a second essay, "What Is Real?," to the slim volume. Written in 1960, this essay contains the dramatic statement: "I have therefore no hesitation in declaring quite bluntly that the acceptance of a really existing material world, as the explanation of the fact that we all find in the end that we are empirically in the same environment, is mystical and metaphysical." In other words, nothing is real. Although Schrödinger reached this conclusion on metaphysical grounds, it resonated strongly with a standard interpretation of the implications of the second quantum revolution, in which he participated in 1926. Schrödinger was pulled back from his philosophical musings and the quiet life in Zürich to become one of the leading participants in this revolution by the discovery of a new way—the right way—to count photons.

Quantum statistics

The discovery came from the Indian physicist Satyendra Nath Bose, via Albert Einstein. Working in Dacca, far away from the European centres where quantum physics was being developed, but keeping in touch by reading the scientific papers being published by the quantum pioneers, Bose came up with his big idea in 1924, at the age of thirty. He found that by using a new kind of statistics to count photons he could derive the Planck black body law entirely from the description of the radiation in a cavity as a quantum "gas" of particles, without invoking waves at all. After the English-language version of a paper reporting his discovery was rejected by the *Philosophical Magazine*, he realized that such a claim might not be taken seriously coming from an unknown Indian researcher, and sent it to Einstein, requesting him to pass it on for publication if he liked it. Einstein was so impressed that he translated the paper into German himself and sent it to the prestigious journal *Zeitschrift für Physik*, where it appeared in

August that year. It is no coincidence that the name "photon" was coined to refer to a particle of light just two years later.

Einstein himself then developed the idea further, applying it to describe the behaviour of any hypothetical collection of atoms—gas or liquid—obeying the same rules. Those rules became known as Bose-Einstein statistics, and they apply to all the quantum entities, such as photons, which are associated with forces (in this case, the electromagnetic force). Such particles are known as bosons. The rules that apply to what we think of as material particles in the everyday sense (things like electrons) became known as Fermi-Dirac statistics, after two of the other quantum pioneers, and the particles are known as fermions. The key distinction is that any number of identical bosons can exist in the same quantum state, but no two identical fermions can exist in the same quantum state—that is, with identical quantum properties. Fermions are also "conserved" in interactions—you cannot increase the number of electrons in the Universe—but bosons can be manufactured indefinitely if there is a source of energy, which is what happens when you switch a light on.

The discovery of Bose-Einstein statistics, and the implications, were hot topics among the cognoscenti at the Innsbruck meeting Schrödinger attended in September 1924. The contact Schrödinger had with Einstein and Planck at that meeting set him thinking about a new line of work on quantum statistics, gas theory, entropy, and statistical mechanics. He developed this thinking during 1925, corresponding with Einstein on the subject throughout that year. The correspondence shows that at first Schrödinger thought that Bose had made only a minor tweak to Planck's calculation, and it was only through Einstein's explanation that he realized the fundamental nature of Bose's contribution. The Einstein connection then pointed Schrödinger towards his masterwork.

In 1924, the French physicist Louis de Broglie (1892–1987) had suggested in his PhD thesis that just as light, traditionally regarded as a wave, behaved in some circumstances like a stream of particles, so electrons, previously regarded as particles, could in some circumstances behave as waves. De Broglie's supervisor, Paul Langevin, was so nonplussed by this that he asked Einstein whether or not to approve the thesis; Einstein said the idea was sound,[4] de Broglie got his doctorate, and the work (of which more in Chapter 7) was also published in the journal *Annales de physique* at the beginning of 1925. Somehow, Schrödinger remained blissfully ignorant of all this until he saw in one of Einstein's papers a reference to de Broglie's idea with the comment that "one was dealing with more than a formal analogy"—in other words, Einstein thought the waves were real. Schrödinger still didn't realize that de Broglie's work had been published in the *Annales de physique*, which he could have found in the library at the university, but he obtained a copy of the thesis almost exactly a year after de Broglie had presented it in Paris. On 3 November 1925, Schrödinger wrote to Einstein: "A few days ago I read with the greatest interest the ingenious thesis of Louis de Broglie, which I finally got hold of. Because of it, [your] work has also become completely clear to me for the first time." After mentioning that there was some connection between de Broglie's ideas and the short paper he had himself published in 1922, Schrödinger went on: "Naturally, de Broglie's consideration within his grand theory is altogether of far greater value than my single statement which, at first, I did not know what to make of."

Within a few weeks of reading de Broglie's thesis, Schrödinger developed a complete, self-consistent theory of the quantum world, based on the idea of waves. But a few months earlier, the German Werner Heisenberg (1901–76) had developed a complete, self-consistent theory of the quantum world, based on the idea of particles. What was going on? The twin discoveries would spark debate that continues to the present day, and would profoundly affect Schrödinger; but before I pick up the thread of his story, I need to digress to tell the story of Heisenberg's discovery of what became known as matrix mechanics. If you already know that story, and in particular if you have read my book *In Search of Schrödinger's Cat*, feel free to skip straight on to Chapter 7. If not, just turn the page.

Chapter Six

Matrix Mechanics

When Werner Heisenberg made the discovery of matrix mechanics, he was even younger than the year of his birth might suggest. As he had been born on 5 December 1901, he was still only twenty-three in the spring and summer of 1925, at the beginning of his career in research. But he had already shown signs of a precocious talent.

Heisenberg was among the first generation of physicists to be brought up on quantum ideas. Between 1920 and 1923 he studied physics and mathematics, first with Arnold Sommerfeld and Wilhelm Wien in Munich, then with Max Born (1882–1970) in Göttingen. He was keenly interested in Niels Bohr's ideas about the behaviour of atoms, and in 1922, although still a student, thanks to Sommerfeld's recommendation he was allowed to attend a major gathering in Göttingen known as the "Bohr Festival."[1] Here he met Bohr himself and heard him give a series of lectures on quantum physics. In his book *Physics and Beyond*, Heisenberg later described the situation in quantum physics at that time as a "peculiar mixture of incomprehensible mumbo jumbo and empirical success." But this "naturally exerted a great fascination." Although he was appointed as a *Privatdozent* in Göttingen in 1924, between Easter that year and spring 1925 Heisenberg had the benefit of working with Bohr in Copenhagen, thanks to a Rockefeller fellowship.

Half-truths

The work that first made Sommerfeld take notice of Heisenberg was carried out when the young man was still a student. By 1920, physicists had become used to the idea of describing the quantum state of a system (such as an atom) in terms of quantum numbers, and it was almost an item of Holy Writ that these always had to be whole numbers—integers such as 1, 2, 3 . . . But Heisenberg realized that certain puzzling features of atomic spectra could be explained if the calculations included half-integer quantum numbers, such as ½, 3/2, 5/2 . . . Sommerfeld was not impressed, and Heisenberg's friend Wolfgang Pauli (1900–58) "suggested that I would soon have to introduce quarters and eighths as well, until finally the whole quantum theory would crumble to dust in my capable hands." So Heisenberg did not pursue the idea. But a few months later an older, established physicist, Alfred Landé (1888–1976), hit upon the same idea and published it.

It turned out that, far from marking the crumbling away of quantum theory, the concept of half-integer quantum numbers was a key to understanding the quantum world, and also that there would be no need for the quarters, eighths, and so on that Pauli had feared. This is best understood in terms of a quantum property known as spin. The spin of an entity such as an electron can be thought of as an arrow which has a certain size but can only point in one of two directions, up or down. Electrons which have opposite spin do not count as identical particles, but electrons with the same spin do count as identical particles, and this affects the way they behave in atoms. But it is best to try to avoid the automatic image that the term "spin" conjures up; things like electrons do not behave like spinning tops or twirling ice-skaters, and quantum spin is a purely quantum property that has no analogy in the everyday world. It is an unfortunate choice of terminology that we are stuck with.

Whatever name it goes by, though, quantum spin is the key to understanding quantum statistics. Objects that have integer or zero spin, such as photons, obey the rules of Bose-Einstein statistics. Objects that have half-integer spin, such as electrons, obey the rules of Fermi-Dirac statistics. But this realization still lay in the future when Heisenberg returned to Göttingen in April 1925 to take up his duties for the summer term.

What you see is what you get

Like many physicists at the time, Heisenberg was puzzling over the nature of electron orbits, the way electrons "jump" between orbits, and how this jumping produces the lines seen in atomic spectra. He was bogged down in a morass of mathematics when, late in May 1925, he was struck by an attack of hay fever so severe that he had to ask his professor, Max Born, for leave of absence to recover. He was granted a two-week break, and on 7 June went straight to the rocky island of Heligoland, far from any sources of pollen.

Heligoland is a tiny island, less than a square mile in area and rising only about 60 metres above the sea, located in the corner of the North Sea known as the German Bight. Because of its location, ownership of the island changed many times until 1714, when it was taken over by Denmark. In 1807, Heligoland was captured by the British during the Napoleonic Wars, and they held on to it until 1890, when it was swapped with Germany for the African island of Zanzibar. When Heisenberg arrived there, after a three-hour journey by ship from Cuxhaven, at the mouth of the Elbe, it was a fading seaside spa resort. "I must have looked quite a sight," he tells us, "with my swollen face; in any case, my landlady took one look at me, concluded that I had been in a fight and promised to nurse me through the aftereffects."[2] But no nursing was required, as the clean air quickly restored him to full fitness, and in between long walks and long swims, with no distractions "I made much swifter progress than I would have done in Göttingen."

Apart from the lack of any distractions on the island, the reason why Heisenberg made such swift progress was that he tried a new way of tackling the problem of quantum jumps. Somebody in the group in Göttingen—nobody could later recall exactly which of them came up with the idea, but the most likely candidate is Pauli—had pointed out that there was no way of knowing what was happening to an atom, or any other quantum entity, when it was not being measured. You could make a measurement which showed the atom to be in a certain quantum state, then make another measurement which showed it to be in another quantum state, but you had no way of telling what had actually happened to the atom in between those measurements. The idea grew up among the group that the only reality that could be described by scientific measurement was the reality of the measurements themselves, and that a physical theory should only be concerned with things that can actually be observed by experiments. In other words, what you see is what you get; neither more nor less. Heisenberg had been a bit dubious about this idea at first—it smacked too much of the philosophical debate about whether a tree falling in a forest makes a sound if there is nobody there to hear it—but he decided to see how such a theory might be developed, and the pieces quickly fell into place.

The crucial thing about observations of quantum systems is that each observation deals with two states at a time. Measuring the energy of a particular line in an atomic spectrum, for example, tells us about the relationship between the two quantum states involved in the process of absorbing or emitting a photon. So Heisenberg began working with a mathematical description of the relationship between pairs of observable quantum states. In the process, he found that he had to work with a particular kind of mathematical entity, a bit like a table of numbers, that described each quantum state.

I've been writing about quantum physics for more than thirty years, and in all that time I've never been able to come up with a better analogy for these mathematical entities than that of a chess board with pieces arranged on it. A chess board is a two-dimensional array of sixty-four squares, and each square can be identified by a letter-number combination, starting with a1 and proceeding through a2, a3, and so on all the way up to h8. The "state" of a chess game can be described by an additional letter to tell you which squares are occupied by which pieces—for example, Qc7 would mean that there is a queen on the square c7 (for simplicity, I'll ignore the difference between black and white pieces). Heisenberg used arrays of numbers not unlike this to describe the quantum state of a system, and worked out the rules for describing the way quantum systems interact to change their states—in effect, multiplying the arrays of numbers together, and performing other mathematical manipulations. As the mathematical description began to fall into place, he decided to apply a crucial test by calculating whether the law of conservation of energy was preserved in his calculations:

> When the first terms seemed to accord with the energy principle, I became rather excited, and I began to make countless arithmetical errors. As a result, it was almost three o'clock in the morning before the final result of my computation lay before me. The energy principle had held for all the terms, and I could no longer doubt the mathematical consistency and coherence of the kind of quantum mechanics to which my calculations pointed. At first I was deeply alarmed. I had the feeling that, through the surface of atomic phenomena, I was looking at a strangely beautiful interior, and felt almost giddy at the thought that I now had to probe this wealth of mathematical structures nature had so generously spread out before me.

But he would not have to carry out the probing alone. When Heisenberg got back to Göttingen and showed his working to Born, Born soon recognized the tables of numbers as examples of a kind of mathematical entity known to mathematicians (but to very few physicists in 1925) as matrices. A paper by Heisenberg announcing his discovery was sent off for publication in *Zeitschrift für Physik*, and Heisenberg himself went off in the summer of 1925 to give lectures in Leiden and Cambridge. Although he did not mention his breakthrough in the lectures, he did discuss it privately; and while he was away, Born and his junior colleague Pascual Jordan (1902–80) developed the theory further using the language of matrices, establishing what became known as matrix mechanics. As Heisenberg commented to Pauli, "It seems as if the electrons will no more move on orbits."

Matrices don't commute

The task of the Göttingen team was made much easier by the stroke of good fortune that Born was one of the few physicists of the time already familiar with matrices. He had had an unusually broad education, having first studied at Breslau, where his father was Professor of Anatomy, and also at Heidelberg, Zürich, and Göttingen, where he concentrated initially on mathematics rather than physics. He had learned about matrices in Breslau. A paper by Born and Jordan extending Heisenberg's work was sent off for publication just two months after Heisenberg's breakthrough paper, and before the end of 1925—before Schrödinger had completed his first paper on what would become known as wave mechanics—Heisenberg, Born, and Jordan had together completed a third paper on matrix mechanics. The key contribution made by Born and Jordan was to stress the central importance of the fact that when you multiply matrices together the answer you get depends on the order in which you do the multiplication. In other words, $a \times b$ is not the same as $b \times a$. This had been implicit, but not spelled out, in Heisenberg's original paper.

In the language of mathematics, matrices do not commute. And using bold letters to denote matrices, with $\mathbf{p}$ and $\mathbf{q}$ representing the quantum equivalents of momentum and position, respectively, Born and Jordan found that

$$\mathbf{pq} - \mathbf{qp} = h/2\pi i$$

where h is Planck's Constant and i is the square root of -1. This relationship is so important that it has become known as the fundamental equation of matrix mechanics, and it is engraved on Born's tombstone. It is an important point, valid throughout quantum mechanics, that if the value of h were actually zero, the equations would boil down to the equations of classical (Newtonian) mechanics; in this particular case, we would have $\mathbf{pq} = \mathbf{qp}$.

But the Göttingen team were not the only ones who had been busy in the second half of 1925. In July that year, when he was in Cambridge, Heisenberg had discussed his work with the physicist Ralph Fowler (1889–1944), and on his return to Göttingen he sent Fowler a copy of his paper, which arrived in Cambridge in August. Fowler passed the paper on to his research student Paul Dirac (1902–84), who was just eight months younger than Heisenberg. Like Born and Jordan, Dirac realized the fundamental importance of the non-commutativity of the variables in quantum mechanics—the fact that matrices do not commute—and working entirely independently, with no knowledge of the work under way in Göttingen, he reworked the entire theory using an elegant branch of mathematics developed in the nineteenth century by the Irish mathematician William Hamilton (1805–65). A copy of the resulting paper was sent to the Göttingen team, and Born later described it as "one of the greatest surprises of my scientific life. For the name Dirac was completely unknown to me, the author appeared to be a youngster, yet everything was perfect in its way and admirable." The author, indeed, was still a student! Dirac's PhD, which he gained in 1926 for a thesis simply titled "Quantum Mechanics," was the first ever awarded for quantum mechanics.

Dirac was undoubtedly the greatest genius of all the people involved in the discovery of quantum mechanics, and also "the strangest man," in the words of his biographer Graham Farmelo, almost certainly because he suffered from a form of autism. Dirac's paper, which also took full account of the need for half-integer quantum numbers, was published in the *Proceedings of the Royal Society* in December 1925. Heisenberg wrote to Dirac: "I have read your extraordinarily beautiful paper on quantum mechanics with the greatest interest, and there can be no doubt that all your results are correct . . . [the paper is] really better written and more concentrated than our attempts here."

So Schrödinger was actually the third person to come up with a complete, self-consistent theory of quantum mechanics. Before long, their achievements would be recognized by the Nobel Committee—but with one scandalous exception.

Justice isn't always done

It soon became clear, as I shall explain in the next chapter, how the work of Heisenberg, Born, Jordan, Dirac, and Schrödinger could be combined to complete the quantum revolution. Armed with the new equations, physicists found that problems that had seemed intractable fell like toppling dominoes. Much later, in his book *Directions in Physics*, Dirac wrote:

> It was a game, a very interesting game one could play. Whenever one solved one of the little problems, one could write a paper about it. It was very easy in those days for any second-rate physicist to do first-rate work. There has not been such a glorious time since then. It is very difficult now for a first-rate physicist to do second-rate work.

By 1928, the various people who had discovered the rules of this new game were already being nominated for Nobel Prizes. In their wisdom the Nobel Committee found one cunning way to share the glory, but also made one glaring omission.

There is a rule that a single Nobel Prize cannot be shared by more than three people, so a more subtle way had to be found to honour all the participants in the second quantum revolution. The solution the committee came up with was to hold the 1932 physics prize over until 1933, and then to award the 1932 prize to Heisenberg, and the 1933 prize jointly to Schrödinger and Dirac, so that they could all be honoured together at the same ceremony. This raises two puzzles. Why weren't Born and Jordan honoured? And if the honours were only going to Heisenberg, Schrödinger, and Dirac, why not let them share a single prize?

We may never know for sure, but the most likely explanation is that by the beginning of 1933 the committee had decided to award the 1932 prize to Heisenberg, Born, and Jordan, and the 1933 prize to Schrödinger and Dirac. But at the beginning of May 1933 Jordan joined the Nazi Party, just at the time Hitler was coming to power in Germany. Unwilling to be seen to endorse an avowed supporter of Hitler's activities, the committee removed both Born and Jordan from consideration, since it would be impossible to disentangle their joint work and award a prize to either of them alone.

The result was embarrassment for Heisenberg, and what Born perceived as a humiliation for himself; he was, after all, the senior member of the Göttingen team. Heisenberg wrote to Born expressing his bad feelings at receiving the prize on his own for "work done in Göttingen in collaboration—you, Jordan and I." Born was bitter about his omission for decades. In 1953 he wrote to Einstein: "In those days [Heisenberg] actually had no idea what a matrix was [until I told him]. It was he who reaped all the rewards of our work together, such as the Nobel Prize." He also commented: "The fact that I did not receive the Nobel Prize in 1932 together with Heisenberg hurt me very much at the time, in spite of a kind letter from Heisenberg." When Born finally did receive a Nobel Prize, in 1954 at the age of seventy-one, nobody was more relieved than Heisenberg. But even this award had a sting in its tail.

The award came for Born's second major contribution to the quantum revolution, which involved the idea that the outcome of events in the quantum world depends on chance and probability—in effect, on the roll of dice. This idea, as I shall explain, was anathema to a few physicists, notably Einstein and Schrödinger; it was the trigger for Schrödinger's famous cat "experiment." But it became the standard way that most physicists thought about the quantum world from the late 1920s onward. Because many of the associated ideas were thrashed out during scientific gatherings at Niels Bohr's Institute in Copenhagen, this way of looking at the quantum world became known as the Copenhagen Interpretation—or, as Born grumbled to Einstein, the "Copenhagen school which today lends its name almost everywhere to the line of thinking

I originated." Born exaggerates slightly, but his ideas were seen as the key to incorporating Schrödinger's wave mechanics into the broader picture—even though Schrödinger himself didn't like that picture.

Time to get back to Schrödinger's work in Zürich at the end of 1925.

Chapter Seven

Schrödinger and the Second Quantum Revolution

Early in November 1925, Peter Debye, at the ETH, asked Schrödinger to prepare a talk for the Zürich physicists about de Broglie's work, which he had read in *Annales de physique*. This was part of a regular series of informal colloquia hosted alternately by the ETH and the university, each attracting an audience of maybe a dozen or two dozen people. The date on which this particular colloquium took place has not been recorded, but it must have been in late November or early December, before the end of the academic term. In a talk he gave to the American Physical Society in 1976,[1] the Swiss physicist Felix Bloch (1905–83), who was a student at the ETH in 1925, recalled that

> Schrödinger gave a beautifully clear account of how de Broglie associated a wave with a particle and how he could obtain the quantization rules of Niels Bohr and [Arnold]

> Sommerfeld by demanding that an integer number of waves should be fitted along a stationary orbit. When he had finished, Debye casually remarked that he thought this way of talking was rather childish. As a student of Sommerfeld he had learned that, to deal properly with waves, one had to have a wave equation. It sounded quite trivial and did not seem to make a great impression [on the group], but Schrödinger evidently thought a bit more about the idea afterwards.

Schrödinger's first thought was to attempt to find a wave equation, moving on from de Broglie's work, that would describe the behaviour of an electron in the simplest atom, hydrogen. He naturally included in his calculations allowance for the effects described by the special theory of relativity, deriving, probably early in December 1925, what became known as the relativistic hydrogen equation. Unfortunately, it didn't work. The predictions of the relativistic equation did not match up with observations of real atoms. We now know that this is because Schrödinger did not allow for the quantum spin of the electron, which is hardly surprising, since the idea of spin had not been introduced into quantum mechanics at that time. But it is particularly worth taking note of this false start, since it highlights the deep and muddy waters in which quantum physicists go swimming—for you need to take account of spin, a property usually associated with particles, in order to derive a wave equation for the electron!

But Schrödinger wasn't stymied for long. With the Christmas break coming up, he had an opportunity to get away from Zürich and think things over in the clean air of Arosa. Inspiration did not, however, come solely from the fresh air and mountain views.

Science and sensuality

Although Schrödinger had many affairs with women, these were seldom, if ever, casual relationships. Judging from his diaries, love was more important to him than sex, although often sex naturally had its place in a loving relationship. He was often in love—or convinced himself that he was in love—and when he was in love, by and large life was good and his scientific creativity benefited. Which is one reason why this aspect of Schrödinger's private life cannot be ignored, even in a scientific biography (there is another reason, too, which I shall come to shortly). Even the science historian Abraham Pais, hardly known for his prurience, felt it necessary, when trying to explain in his book *Inward Bound* the baffling success of thirty-eight-year-old Schrödinger in December 1925, to refer to "the remark once made to me by Hermann Weyl that Schrödinger did his great work during a late erotic outburst in his life." Weyl, of course, was Anny Schrödinger's lover, so presumably knew what he was talking about.

The point is that Schrödinger was not alone in Arosa. For the previous two Christmases he had been there with Anny; but this time he was accompanied by an old girlfriend from Vienna. We don't know who she was, because although Schrödinger's diaries are usually quite explicit on such matters, the relevant volume is missing. Whoever she was, though, she seems to have triggered a burst of creative activity which carried Schrödinger right through 1926, producing six major scientific papers on what became known as wave mechanics. But it all began with what looks at first like a backward step. To put it in context, we need to take a quick look at what it was that de Broglie had actually done.

Louis de Broglie very nearly didn't become a physicist. As the younger son of a French noble family (he later inherited the title "Prince" from his elder brother, Maurice), he initially studied history and was expected to enter the diplomatic service. But Maurice, who was seventeen years older than Louis, had become a respected experimental physicist by the time Louis entered the Sorbonne in 1910—just at the time the world of science was becoming excited about quantum physics. Encouraged by Maurice, Louis switched to physics, but his education was interrupted by military service in the First World War, when he worked in radio communications at the Eiffel Tower. So it was not until 1923 that he published his first papers on the nature of light quanta and developed the work that formed the basis of his doctoral thesis, and not until late 1924 that he was awarded his PhD (after Einstein's intervention!), by which time he was already thirty-two years old. Hardly surprisingly, de Broglie made no further great contributions to physics; the surprise is that the even older Erwin Schrödinger was the person who picked up the ball and ran with it.

De Broglie had started out from the two equations that Einstein had derived for light quanta:

$$E = h\nu$$

and

$$p = h\nu/c$$

Since the wavelength (λ) of a wave is related to its frequency (ν) by the equation

$$\lambda = c/\nu$$

a very simple substitution then told him that

$$p\lambda = h$$

—or, in plain English, that the momentum of a quantum entity multiplied by its wavelength is equal to Planck's Constant. And this applies, in principle, to any object—it tells us that there is a momentum associated with light waves, but also that there is a wave associated with electrons and other particles. The equation also makes it clear why such effects are not observed in the everyday world: because the momentum of anything we can see or touch so big compared with Planck's Constant, the waviness associated with such an object is far too small to be noticed.

After his failure with the relativistic hydrogen equation, Schrödinger went back to basics. Starting with the standard wave equation of classical mechanics, he used the relationship discovered by de Broglie to convert the wavelengths to momenta, and came up with a very simple wave equation for the electron, similar to the wave equation for light and other electromagnetic waves discovered by Maxwell in the nineteenth century. The derivation is one of those ideas that, like de Broglie's discovery, seems almost embarrassingly simple once someone has thought of it, provoking the reaction "Why on earth didn't I think of that?," but was far from obvious until it was pointed out. Perhaps Schrödinger also thought it was embarrassingly simple, because before he published his breakthrough he came up with two other, more complicated (and therefore more impressive) ways of deriving his famous wave equation to share with the world of science. And, unlike Schrödinger's relativistic equation from a few weeks earlier, this one predicted the right values for the quantum numbers determined by experiment. Einstein had been right to say that de Broglie's equation represented "more than a formal analogy."

The bizarre thing was that this treatment, ignoring the implications of the special theory of relativity, worked, when it had no right to do so. With hindsight, we can see that the consequences of ignoring the special theory in effect cancelled out the consequences of ignoring spin. The "right" equation for describing quantum waves can indeed be derived properly by incorporating both relativity and spin, but by luck it can also be derived by ignoring both of them. On such luck, Nobel Prizes sometimes depend.

Things moved swiftly after Christmas 1925. At the beginning of the next term in Zürich, Schrödinger gave another colloquium, in which, Bloch recalled in 1976, he began with the words: "My colleague Debye suggested that one should have a wave equation [for the electron in the hydrogen atom]; well, I have found one." This was only true up to a point; a great deal of hard work was needed to produce a complete mathematical description of the hydrogen atom in terms of the "nonrelativistic hydrogen equation." But with the help of his colleagues in Zürich, Schrödinger completed his first paper on wave mechanics and sent it off to *Annalen der Physik*, where it arrived on 27 January, less than a month after the breakthrough in Arosa, and was published on 13 March 1926. By then the same journal had already received a second paper from Schrödinger, developing the idea further, which was published on 6 April, followed in swift succession by four more papers, the last published on 5 September. As if that were not enough, Schrödinger also wrote a comprehensive overview titled "An Undulatory Theory of the Mechanics of Atoms and Molecules," which was finished on 6 September and published in English in the journal *Physical Review* in December 1926. Wave mechanics was essentially complete, less than a year after the failure with the relativistic hydrogen equation. The key papers were collected in a single volume published the following year in German, and in 1928 in English. It was an astonishing creative outburst, unequalled in science by anyone of a comparable age to Schrödinger in 1926, and arguably surpassed only by the young Albert Einstein's productivity in his *annus mirabilis* of 1905, when he made key contributions to several different areas of science. Einstein himself was impressed; in May 1926 he wrote to his friend Michele Besso, "Schrödinger has come out with a pair of wonderful papers on the quantum rules."[2]

Even Schrödinger, though, was surprised by the outcome of his investigations. In the introduction to the collected edition of his papers on wave mechanics, he wrote:

> A young lady friend recently remarked to the author: "When you began this work you had no idea that anything so clever would come out of it, had you?" This remark, with which I wholeheartedly agreed (with due qualification of the flattering adjective), may serve to call attention to the fact that the papers now combined in one volume were originally written one by one at different times. The results of the later sections were largely unknown to the writer of the earlier ones. Consequently, the material has unfortunately not always been set forth in as orderly and as systematic [a] way as might be desired, and further, the papers exhibit a gradual development of ideas.

The reference to a "young lady friend" highlights the continuing importance of both eroticism and science in the year of Schrödinger's "late erotic outburst," although which (if either) came first it is impossible to say. She was fourteen-year-old Itha Junger, whom Schrödinger was coaching in mathematics—among other things.

Itha was one of a pair of non-identical twin sisters whose mother was an acquaintance of Anny Schrödinger. For much of July 1926, at the end of the academic year, Erwin had been away from Zürich, travelling to other research centres, including Munich and Berlin, spreading the news of his breakthrough and raising his profile considerably in the German-speaking scientific world. In his absence, Anny learned from their mother that the twins (christened Itha and Roswitha, but known as Ithi and Withi) had failed to achieve the required standard in mathematics at their convent school; Itha had done particularly badly, and might have to be held back for a year, either splitting the twins up or forcing Roswitha to be held back with her. Anny suggested that Erwin would be the ideal person to give them some special tuition. He returned shortly before the twins' fourteenth birthday in August, and eagerly took up the challenge. The consequences were described to Walter Moore by Itha in a series of interviews in 1985.

The maths lessons were a great success, with most of their tutor's attention naturally being devoted to Itha, and both girls achieved the standard required to move on with their classmates when the next school term began. But as well as the maths, the lessons included "a fair amount of petting and cuddling," and Schrödinger soon convinced himself that he was in love with Ithi ("whatever love is," as Prince Charles once said). He talked to her about his scientific work and about his religious beliefs, wrote poetry for her, spent a skiing holiday with the two girls and their mother over Christmas 1927, and set out on a long campaign of what would now be described as grooming. Of course, the head of the young convent-school girl was turned by all the attention, and in due course she fell in love with him. But he was patient. It wasn't until she was sixteen that he went into Ithi's room in the middle of the night (during another skiing holiday) and told her how much he loved her; and not until just after her seventeenth birthday, in August 1929, that the relationship was consummated. The affair continued into the 1930s, with Schrödinger at one point seriously considering divorcing Anny to marry Ithi, and forms a backdrop which cannot be ignored to Schrödinger's scientific life in the years following his discovery of wave mechanics. But the scientific world knew nothing of this when it began to take wave mechanics seriously and developed it into a complete description of the quantum world. Unfortunately for Schrödinger, though, the way his idea was developed was not at all to his taste.

Riding the wave

Schrödinger's waves were a classic example of a continuous process; Heisenberg's matrices provided a classic description of a discontinuous process, which Schrödinger found repugnant. In his paper "On the Relation of the Heisenberg–Born–Jordan Quantum Mechanics to Mine," published in *Annalen der Physik* in May 1926, he noted:

> My theory was inspired by L. de Broglie and by brief but infinitely far-seeing remarks of A. Einstein . . . I was

> absolutely unaware of any genetic relationship with Heisenberg. I naturally knew about his theory, but because of the to me very difficult seeming methods . . . I felt deterred by it, if not to say repelled.

So it came as something of a surprise, not to say a shock, when he found that not only did the two theories give the same (correct) answers when applied to the same problems in atomic physics, but that they were mathematically equivalent. Matrix mechanics could be derived from wave mechanics by taking the variables corresponding to position and momentum in Schrödinger's wave equation and replacing them with two of the expressions, known as operators, from Heisenberg's theory. This is a process the mathematicians call "substitution," and it works equally well going the other way, from matrix mechanics to wave mechanics.

Schrödinger wasn't the only person to spot the links between the two theories. Pauli had also noticed the connection, and mentioned it in a letter to Jordan in April 1926, before he had seen Schrödinger's paper. And the American Carl Eckart (1902–73), based in Pasadena, wrote two papers on the topic in May and June 1926, before copies of the issue of the *Annalen* containing Schrödinger's paper had reached California. Eckart's achievement, in the words he used in the first of those two papers, was "the inclusion of the results of Schrödinger in a single calculus with those of [Heisenberg, Born, and Jordan]. . . . This would seem to be the greatest support which either of the two dissimilar theories have thus far received." But the definitive last word on the subject came from the boy wonder, Paul Dirac.

Schrödinger, Pauli, and Eckart had each shown empirically that matrix mechanics and wave mechanics were equivalent to one another at the level of making substitutions. But none of them had been able to say why this should be the case. Dirac developed yet another way of looking at the quantum world, which he called transformation theory, and proved (using some hairy mathematics) that all versions of quantum mechanics were contained within this overarching theory. His paper describing the new theory was received by the Royal Society, for publication in their *Proceedings*, in December 1926; Jordan did some similar work at the same time, although without quite scaling the heights Dirac reached.

Without going into the hairy mathematics, the best way to understand Dirac's achievement is with the aid of an analogy which Heinz Pagels used in his book *The Cosmic Code*. He points out that a tree may be described in two (or more) different languages, say, English and Arabic. The two descriptions may look utterly different—in this example, they do not even use the same alphabet. But they both describe the same thing, and one description can be transformed into the other using a dictionary and the rules of grammar. "That different representations are subject to laws of transformation is a profound idea," says Pagels. "Invariants establish the true structure of an object."

Transformation theory is the complete theory of quantum mechanics. But few ordinary physicists in the 1920s (and few since, for that matter) bothered about that. They didn't like the hairy mathematics, and most of them, like Heisenberg, hadn't even known what a matrix was before 1926. What they seized on was the fact that if matrix mechanics and wave mechanics were equivalent, when it came to solving practical problems, then you could use whichever one you liked when confronted by those problems. And the one people liked was wave mechanics, since they all knew (or thought they knew) how waves behaved. Which was great news for Schrödinger—at first.

The problem for Schrödinger was that even if the maths said that matrix mechanics and wave mechanics were equivalent, that didn't give him a physical picture of what was going on inside atoms. How could discontinuous quantum jumping be reconciled with a continuous wave function? He puzzled over the problem during the summer of 1926 (with light relief from this work in the form of tutoring the twins), but was getting nowhere, as a letter to Wilhelm Wien, written on 25 August and now in the Wien Archive in Munich, testifies:

> The photoelectric effect [see Chapter 4 above] . . . offers the greatest conceptual difficulty for the achievement of a classical theory . . . I no longer like to assume with Born that an individual process of this kind is "absolutely random" . . . I no longer believe today that this conception (which I championed so enthusiastically four years ago) accomplishes much.

Schrödinger was particularly concerned about the statistical interpretation at this time because in June 1926 Max Born had come up with a new way of interpreting the wave function. He suggested that the wave function could be used to calculate the probability of a quantum entity such as an electron being at a particular point in space. A defining feature of a wave is that it has a size, or amplitude, which varies from place to place, and Born found that the square of the amplitude of Schrödinger's wave function could be used as a measure of probability.[3] He suggested that particles such as electrons are real entities, but that where you find them depends on the probability amplitudes associated with a ghostly wave. The snag, as far as Schrödinger was concerned, was that this meant that an entity such as an electron did not have a definite path, or trajectory, through space, but might be found anywhere in a certain region of space determined by the probabilities. Instead, Schrödinger favoured the idea that a particle is somehow guided by a field which obeys the wave equation, so that the particle rides the wave, like a surfer. But: "It would depend on the taste of the observer which he now wishes to regard as real, the particle or the guiding field." Muddy waters, indeed! And a clear break with Schrödinger's earlier ideas about the importance of statistical processes in even the most fundamental situations in physics. The irony is that whereas Schrödinger had previously been in a minority when suggesting that the fundamental laws of physics are statistical, from now on he would be in a minority in rejecting the idea that statistics play a vital role in the behaviour of the quantum world.

With these ideas buzzing in his head, Schrödinger went on holiday with Anny in the South Tyrol, then on to Copenhagen at the end of September to discuss quantum physics with Bohr and his colleagues, including Heisenberg, who was working at Bohr's institute at the time. Schrödinger gave a lecture on wave mechanics there on 4 October, but the most important feature of the visit was the opportunity to discuss his ideas, especially his concern about quantum jumps, with Bohr. He had plenty of opportunity to do so, since he was staying at the Bohrs' house. Each man held firmly to his position—just how firmly we can tell from Heisenberg's detailed account of the debate and its aftermath in his book *Physics and Beyond*, where he tells us that:

> Although Bohr was normally most considerate and friendly in his dealings with people, he now struck me as an almost remorseless fanatic, one who was not prepared to make the least concession or grant that he could ever be mistaken.
>
> Schrödinger was equally stubborn.
>
> It is hardly possible to convey just how passionate the discussions were, just how deeply rooted the convictions of each, a fact that marked their every utterance.

Schrödinger's entrenched view was that if there were no laws to describe the motion of an electron during a quantum jump then "the whole idea of quantum jumps is sheer fantasy"; Bohr's firmly held conviction was that this

> does not prove that there are no quantum jumps. It only proves that we cannot imagine them, that the representational concepts with which we describe events in daily life and experiments in classical physics are inadequate when it comes to describing quantum jumps. Nor should we be surprised to find it so, seeing that the processes involved are not the objects of direct experience.

It was in the course of this debate that Schrödinger made his famous comment "if all this damned quantum jumping were really here to stay, I should be sorry I ever got involved with quantum theory."

Although the debate was inconclusive, it left Schrödinger, Bohr, and Heisenberg (still good friends in spite of their intellectual differences) all deeply puzzled about where quantum theory was taking them. It was puzzling over the questions raised during Schrödinger's visit to Copenhagen that led Heisenberg to the next great discovery, essentially the last piece in the jigsaw puzzle of quantum mechanics. It came to him on a winter's night in an attic in Copenhagen; Schrödinger, on an extended visit to the United States (described in the next chapter), would learn nothing of this new idea until he returned to Europe in April 1927.

A quantum of uncertainty

Although convinced that they were right, the Copenhagen scientists had now realized "how difficult it would be to convince even leading physicists that they must abandon all attempts to construct perceptual models of atomic processes." In the months following Schrödinger's visit to Copenhagen, the physical interpretation of quantum physics was the central theme of the discussions between Bohr and Heisenberg, with the other piece of the puzzle that had been found by Max Born in the background. One of the key problems that occupied them over the weeks up to Christmas was how to reconcile either version of the new quantum mechanics with "so simple a problem as the trajectory of an electron in a cloud chamber."

A cloud chamber is a relatively simple device: essentially, a sealed box containing air saturated with water vapour, with a glass window through which to watch, and photograph, what is going on. When a particle such as an electron zips through the chamber, it leaves a trail of condensation behind it, similar to the way a high-flying aircraft produces a condensation trail across the sky. The cloud chamber had been invented in the 1890s by Charles Wilson (1869–1959), who received a Nobel Prize for his work in 1927; but the technique had only been fully developed after 1910, by Patrick Blackett (1897–1974), who also received a Nobel Prize, in 1948. The awards indicate the importance of the cloud chamber to the new physics. This was the nearest anyone could come in the 1920s to seeing an individual electron, and the tracks produced really did look like the effects of fast-moving particles.

Bohr and Heisenberg were puzzled because the very concept of a trajectory did not fit in with the ideas of matrix mechanics (which Heisenberg himself always referred to as quantum mechanics, although today that term embraces wave mechanics as well). Although it is possible in wave theory to have a localized bunch of waves moving together as a so-called wave packet, this would require a beam of matter spread out over a width much greater than the diameter of an electron. Not what was seen in the cloud chamber.

Early in the new year, with their discussions continuing long after midnight for weeks on end, both Bohr and Heisenberg "became utterly exhausted and rather tense." So Heisenberg was more than happy when Bohr decided to go skiing in Norway in February 1927, leaving him to "think about these hopelessly complicated problems undisturbed." He did so in his cosy flat at the top of the building which housed Bohr's institute, looking out over Copenhagen. Back in the spring of 1926, Heisenberg had been asked to give a talk on matrix mechanics at the University of Berlin, and afterwards had had a long discussion with Einstein about the nature of reality and the implications of the new theory. At one point, Einstein had commented: "It is quite wrong to try founding a theory on observable magnitudes alone. In reality the very opposite happens. It is the theory which decides what we can observe." At the time, Heisenberg was completely taken aback by Einstein's argument. But now, nearly a year later, the words suddenly came back to him one night after midnight, when he was wrestling with the puzzle of trajectories. It is the theory which decides what we can observe. Could this be the key? Too excited to sit at his desk any longer, Heisenberg went for a walk through the nearby Faelled Park. It was on that nocturnal walk that he came up with the idea forever associated with his name—quantum uncertainty.

Heisenberg had realized that what is actually observed in a cloud chamber is a trail of water droplets triggered into condensation by the electron. This is not the continuous path of an electron through the chamber, but a series "of discrete and ill-defined spots through which the electron has passed," like a line of dots which we join up to create a trajectory. We don't know what the electron is doing in between those dots, any more than we know what it is doing when it "jumps" between energy levels in an atom. So Heisenberg reasoned that the right question to ask was: "Can quantum mechanics represent the fact that an electron finds itself *approximately* in a given place and that it moves *approximately* with a given velocity, and can we make these approximations so close that they do not cause experimental difficulties?"[4]

Hurrying back to the institute, armed with this insight Heisenberg was quickly able to prove mathematically that everything was consistent provided that the system obeyed a simple rule—what became known as Heisenberg's Uncertainty Principle. In his own words: "The product of the uncertainties in the measured values of the position and momentum . . . cannot be smaller than Planck's Constant."

One slight adjustment has been made subsequently—we now know that the product of the uncertainties has to be less than Planck's Constant divided by 4π. On 23 February, Heisenberg wrote a long letter to Pauli outlining his discovery—essentially, a draft version of the paper that he was preparing for the *Zeitschrift für Physik*. "The path," he said, "only comes into existence through this, that we observe it."

But what does this mean physically? The first key point is that the word "measured" can be removed from Heisenberg's statement. Quantum uncertainty has nothing to do with our ability, or inability, to make precise measurements. It is something that is built into the quantum world, so that an entity such as an electron does not have both a precisely determined momentum (which in effect means velocity) and a precisely determined position. As Heisenberg put it when he published the news of his discovery in the *Zeitschrift* later in 1927, "We cannot know, as a matter of principle, the present in all its details."

The electron itself does not "know" both exactly where it is and exactly where it is going at the same time. But because the constraint (Planck's Constant divided by 4π) is a product of the two uncertainties, either one of them can be specified as precisely as you like, provided there is correspondingly larger uncertainty for the other. The more accurately the velocity of an electron is specified, the less accurately is its position specified; the more accurately its position is specified, the less accurately is its velocity specified. In the cloud chamber, the trajectory of the electron specifies its velocity fairly precisely, but its position could be anywhere along the trajectory. In principle, the same rules apply in the everyday world. But quantum uncertainty does not noticeably affect objects more massive than molecules, because Planck's Constant is so small. The amount of uncertainty in the position of an object is proportional to Planck's Constant divided by the mass of the object, and for everyday objects the uncertainty is absolutely tiny. If we were the size of an electron, though, quantum uncertainty would be common sense.

And this brings probability and statistics back to centre stage. In the *Zeitschrift*, Heisenberg wrote:

> "When we know the present precisely, we can predict the future," is not the conclusion but the assumption. Even in principle we cannot know the present in all detail. For that reason everything observed is a selection from a plenum of possibilities and a limitation on what is possible in the future. As the statistical character of quantum theory is so closely linked to the inexactness of all perceptions, one might be led to the presumption that behind the perceived statistical world there still hides a "real" world in which causality holds. But such speculations seem to us, to say it explicitly, fruitless and senseless.

Many disagreed, not least Schrödinger. Bohr, on his return from Norway, at first thought Heisenberg was barking up the wrong tree, and that his concept of uncertainty was incompatible with Bohr's own developing ideas about the quantum world. But within a few months of fierce debate among the experts, a new perception of the quantum world emerged, and became the received wisdom (though not without dissenters) for more than half a century. It was the consensus known, to Born's intense annoyance, as the "Copenhagen Interpretation."

The Copenhagen consensus

When Bohr got back to Copenhagen from his skiing trip, he brought with him another new idea about the quantum world. It later became known as complementarity, and it lies at the heart of the Copenhagen Interpretation. It is a disarmingly simple idea at first sight, but with deep ramifications.

Bohr suggested that both the wave and the particle descriptions of the quantum world are correct, and that they are complementary aspects of some greater whole. The simplest analogy—no less good for being simple—is with the two sides of a coin. You can see either the head or the tail, but not both at once; they are complementary aspects of the coin. Similarly, if we do an experiment with electrons designed to find waves, we will see waves; but if we do an experiment with electrons designed to find particles, we will see particles. What the electron "really is" is irrelevant, and perhaps beyond human understanding. What matters is what it is like, or how it behaves in particular circumstances.

This led Bohr and Heisenberg into conflict in March 1927, because Heisenberg wanted to do away with the idea of waves altogether. So at first he was less than enthusiastic when Bohr found a simple way to derive the uncertainty relation using wave mechanics. This proved to be very easy, because although a localized group of waves (a wave packet) can indeed behave in some ways like a particle, this can only be achieved if the packet contains many waves with different wavelengths. A single wave has a well-determined momentum, but, of course, it does not have a well-determined position. A wave packet has a well-determined position, but it does not have a well-determined momentum. When the numbers are put in, Heisenberg's uncertainty relation comes out.

Heisenberg finally had to admit that Bohr was right, and included a postscript referring to Bohr's work in his uncertainty paper, which was published at the end of May. On the strength of his work, proving that he was no one-hit wonder, Heisenberg moved to Leipzig as a full professor (the youngest in Germany) at the end of June.

Meanwhile, Bohr struggled to get his own ideas down on paper. He had notorious difficulty with this throughout his career, going through endless drafts and re-drafts in an attempt to find the precise form of words he needed. He was usually helped in the task by an assistant, and in this case that far from easy role fell to Oskar Klein (1894–1977), a Swedish physicist working at Bohr's institute. Draft followed draft, with the term "complementarity" appearing for the first time in a version dated 10 July 1927. Bohr was working against a deadline, to complete his paper in time to present it at a conference to be held in Como that September to mark the centenary of the death of Alessandro Volta (1745–1827), but in typical Bohr fashion he failed to finish on time, and had to make a presentation based on the latest working copy.

The Como meeting was the first opportunity to present the package of ideas that became known as the Copenhagen Interpretation to the leading physicists of the day. But these did not include Schrödinger, who, as I describe in the next chapter, was settling in after moving to Berlin and was too busy to attend, or Einstein, who refused to visit Fascist Italy. The essential elements of the package Bohr described included Schrödinger's wave equation, now interpreted as a "probability wave," Born's statistics, Heisenberg's Uncertainty Principle, complementarity, and—something that would trouble Schrödinger and Einstein deeply—a notion called "the collapse of the wave function." Bohr also stressed that the only reality lay in the observations—that it was meaningless to ask where a quantum entity such as an electron was, or what it was doing, when it was not being observed. The best way to understand how the interpretation works is by looking at the classic puzzle of the quantum world, the double slit experiment. I'll describe this in terms of electrons, but the same interpretation applies to any quantum entity.

According to the Copenhagen Interpretation, when an electron is ejected from an electron "gun" on one side of the experiment it leaves as a particle, and can be detected as a particle. But it immediately dissolves into a probability wave, which travels through both of the holes and interferes with itself to make a pattern of probability on the other side of the holes. At the detector screen, the electron can appear as a particle at any point allowed by the probabilities, but with some places more likely than others, and, crucially, some locations being absolutely forbidden. There is a "collapse of the wave function" at the point where the electron is observed, or measured. It arrives as a particle. But the moment that it is no longer being observed, the probabilities spread out again from that point, so that next time we look for the electron we find it somewhere else, that somewhere being anywhere allowed by the probabilities but, as ever, with some locations more likely than others.

Do not imagine, though, that the probability wave is in some sense a "smeared-out" version of the electron. The electron is only ever seen as a particle—we do not, for example, see half of the electric charge of the electron passing through each of the two holes—but the place where the particle is seen or detected depends on the statistical rules determined by the behaviour of the wave. And do not imagine that because the rules are statistical the Copenhagen Interpretation applies only to the behaviour of thousands, or billions, of separate electrons (or other quantum entities) added together. It applies to each individual electron, just as I have described—and, unlike Bohr and his contemporaries, we have the benefit of modern versions of the double slit experiment carried out by firing electrons literally one at a time through single slit and then double slit, which have confirmed every prediction of his description of quantum reality.

The audience in Como were as baffled by Bohr's presentation of this new idea as most people are when they first encounter it. Few, if any, were convinced. But the following month, in the last week of October 1927, he had a second chance to make his case, and this time he grasped it with both hands. The occasion was the fifth Solvay Congress, one of a series of scientific conferences held in Brussels, established in 1911 under the sponsorship of a wealthy Belgian industrial chemist, Ernest Solvay. It turned out to be the greatest gathering of quantum physicists ever. The grand old men of the first quantum revolution—Planck, Einstein, and Bohr—were all there; but so too were the young men of the second quantum revolution, including de Broglie, Dirac, Heisenberg, and Pauli. There, too, was Schrödinger, who was at once both a grand old man (as physicists judge age) and a key player in the second quantum revolution. The official title of the conference was "Electrons and Photons," and the invitation sent out to prospective participants stressed that it would be "devoted to the new quantum mechanics and to questions connected with it."

What happened at the meeting is best summed up by the comment of physicist Paul Ehrenfest (1880–1933) in a letter he wrote just after the conference: "BOHR towering completely over everybody. At first not understood at all . . . then step by step defeating everybody."[5] A slightly more thoughtful assessment of the outcome of the conference came from Heisenberg in 1963: "The most important success of the Brussels meeting was that we could see that against any objections, against any attempts to disprove the theory, we could get along with it. We could get anything clear by using the old words and limiting them by the uncertainty relations and still get a completely consistent picture."[6] Either way, the second quantum revolution was complete and the Copenhagen Interpretation held centre stage. As Schrödinger began his new life as a professor in Berlin, this was not at all to his liking.

Chapter Eight

The Big Time in Berlin

Schrödinger's discovery of wave mechanics came at the perfect time for his career. Max Planck, who had been Professor of Theoretical Physics in Berlin since 1892, was approaching his seventieth birthday, the age at which he would be required to retire. In the summer of 1926, exactly at the time Schrödinger's ideas were being published to widespread acclaim, a committee was set up to consider the appointment of Planck's successor. The Berlin chair was the top job in theoretical physics in mainland Europe, and the committee could pick and choose from the best people; there was no question of having to advertise the post. Einstein was out of consideration, though. He was already in Berlin, where he had a special professorship which carried no teaching duties at all. Heisenberg was a contender, but in the summer of 1926 he was still only twenty-four, ten years younger than even Planck had been at the time of his appointment; it was felt that the opportunity had come too soon for him, in spite of his proven ability.

The committee finally came up with a short list of two—Max Born and Erwin Schrödinger. By December 1926, where we pick up the story of Schrödinger's life following his own *annus mirabilis*, Erwin knew that he was being considered for the post; but nothing was decided when he set out with Anny on a working trip to the United States.

Making waves in America

The trip stemmed from an invitation made by the University of Wisconsin, which offered Schrödinger $2,500 (including expenses) to give a series of lectures on wave mechanics. He was initially reluctant to take up the offer, not least because it would mean being away from Zürich for the Christmas vacation (was he thinking of Ithi?), but was persuaded by Anny that it was too good an opportunity to miss. The couple left on 18 December, travelling by train and ship (spending Christmas on board), and arriving in New York for the new year. A memoir written by Anny, now in the Schrödinger Archive in Vienna, gives us a taste of their experiences in the United States, and an insight into Schrödinger's character. Always a lover of the great outdoors, he hated New York so much that he threatened to go home on the next ship. After a single night in the city they moved on by train to Madison, via Chicago (where he was terrified of being shot at by gangsters), and here things calmed down: Schrödinger regarded Madison as a proper city in the European mould. His lectures went down so well that he was offered a chair at the university; he had no intention of moving to America, a country so uncivilized that alcohol was prohibited, but was able to decline tactfully by letting it be known that he was in the running for the Berlin post.

Having got over his initial impressions of New York, Schrödinger was eager to spread the word about wave mechanics, and gave lectures at the midwestern universities of Chicago, Iowa, and Minnesota, making do with ginger ale as an accompaniment to the fine meals he was offered as guest of honour. Then it was on to California, which the Schrödingers loved (regretting only that it was populated by Americans rather than by Italians or Spanish), and another lecture, at the California Institute of Technology in Pasadena. The return trip east took in Ann Arbor, lectures at both Harvard and MIT, then Baltimore and Washington. Trailing clouds of glory, Schrödinger was now offered a professorship at Johns Hopkins University on a salary of $10,000 a year; but nothing (especially not anything in America) would tempt him to miss out on the possibility of the Berlin chair.

Fitting in one more lecture at Columbia University, making a total of more than fifty in just under three months, Erwin finally left the New World for home, leaving wave mechanics firmly established in the minds of physicists across the United States as the leading version of quantum mechanics. He and Anny arrived back in Zürich on 10 April 1927, just as the committee in Berlin were making up their minds.

Berlin and Brussels

The choice between Born and Schrödinger had proved difficult, but the report of the committee, now in the Berlin archive, shows that they were swayed by the breadth of

Schrödinger's research, by his "deep and original ideas," and in particular by his "especially daring design through his ingenious idea for the solution of the former particle mechanics by means of a wave mechanics." They also commented on his "superb" style as a lecturer, "marked by simplicity and precision," and mentioned that he had "the charming temperament of a South German." Had he been aware of it, Schrödinger would surely have loved the reference to particle mechanics as a thing of the past; but as an Austrian he might not have thought so highly of being described as a South German.

So the offer went out to Schrödinger at the beginning of the summer term. Although he had keenly wanted the honour of being invited to Berlin, now the ball was in his court Schrödinger hesitated before accepting the offer. He had by now settled happily in Switzerland, valuing the peace and stability which contrasted with much of the rest of Europe at that time, and enjoying the proximity of the mountains. As a place to live, it was almost as good as Austria. The university in Zürich made every effort to keep him; although they could not match the salary offered by Berlin, they proposed giving Schrödinger a joint professorship at the university and the ETH, providing twice as much money, but unfortunately requiring twice as much lecturing.

It wasn't only the university authorities who wanted to keep him. When the students heard the news that he might be leaving, they organized a torchlight procession through the streets to Schrödinger's house—a traditional, but very rare, way for the student body to honour favourite teachers. But while Schrödinger hesitated, the balance was tipped by a remark from Planck himself: "It would make me happy." So the Schrödingers moved to Berlin at the end of the summer term, shortly before Erwin's fortieth birthday, although he would not take up his formal duties until 1 October 1927. His lectures did not begin until 1 November, after he had returned from the Solvay Congress in Brussels, which he saw in a rather different perspective from Heisenberg's.

Schrödinger must have enjoyed one aspect of Louis de Broglie's early contribution to the proceedings, since de Broglie tried to do away with the idea of interpreting Schrödinger's wave function in terms of probabilities. But other aspects of his presentation were less appealing to him. De Broglie offered what became known as the pilot wave model, which says that an electron consists of two linked and physically real entities, a wave and a particle, with the particle riding the wave like a surfer—in marked contrast to Bohr's idea that the electron behaves either as a wave or as a particle, but never as both together. De Broglie's idea received little attention from the delegates, although a version of it was taken up later by some physicists and played a key part, as we shall see, in some of the most exciting developments in physics during the late twentieth century. In 1927, the physicists at the Solvay Congress saw the debate largely in terms of the discussions between Bohr and Heisenberg on one side and Schrödinger and Einstein on the other, with the Copenhagen approach winning out and the more diffident de Broglie being ignored.

One reason why Schrödinger lost the debate was that his ideas about the nature of electrons were too abstract for simple physicists to feel comfortable with. For example, the equation for a single electron involves a wave moving in three dimensions; if a second electron interacts with the first electron, that requires another wave moving in three different dimensions. Mathematicians are used to such ideas, and call the fictitious space in which the waves interact "phase space." But the idea of an extra three dimensions of phase space for every particle seemed, in 1927, far less attractive than the Copenhagen Interpretation. Although Schrödinger expressed the hope that future developments of the theory would lead to a more conventional version involving the four ordinary dimensions of spacetime, Heisenberg stood up and replied: "I see nothing in Mr. Schrödinger's calculations that would justify this hope."

Schrödinger's presentation, though, also included a remark which passed largely unnoticed at the time, but today seems remarkably prescient: "The real system is a composite image of the classical system in all its possible states." More of this later; but hold on to that idea of "the classical system in all its possible states."

Einstein was not one of the main speakers at the conference, but made a comment on Bohr's contribution which raised a subject that would concern those quantum physicists who worried about such things for decades. He pointed out that in the double slit experiment, when Bohr's probability wave arrives at the detector screen there is a particular probability of finding the electron associated with each point on the screen. But as soon as the electron is detected at one point, the probability everywhere else becomes zero—instantly. It is as if a signal of some kind has travelled across the detector screen in an instant—crucially, faster than the speed of light. But according to the theory of relativity, nothing can travel faster than light. The collapse of the wave function seemed to require faster-than-light signalling: what Einstein called "a peculiar instantaneous action at a distance." The puzzle persisted into the 1980s (see Chapter 14).

But very few quantum physicists did worry about such things. The Copenhagen Interpretation gave them a package which worked, and was easy to use in calculations. They didn't care about the philosophical interpretation of what was going on any more than the average car driver cares how an internal combustion engine works. At the other extreme, Dirac wasn't interested in interpretation because he believed that the truth lay in the equations, and it was pointless to ask about their physical meaning.

In the months following the Brussels conference, Heisenberg took up a professorial post in Leipzig, Pauli became a professor at the ETH, and Pascual Jordan succeeded Pauli in Hamburg. Born was already established in Göttingen, while Bohr continued to rule the roost in Copenhagen. Under their influence, the Copenhagen Interpretation thrived, and was put to many practical uses, not least in explaining how atoms join together to make molecules. By 1929, Dirac was able to write in one of his scientific papers, completely accurately, that "the underlying physical laws necessary for the mathematical theory of a large part of physics and the whole of chemistry are thus completely known." It was half a century later that he wrote about how "It was very easy in those days for any second-rate physicist to do first-rate work."[1]

So it was with mixed feelings that Schrödinger returned to Berlin to take up his lecturing duties in November 1927. Wave mechanics was thriving—but not in a form he approved of. The consolation was the presence in Berlin of Einstein, who shared his doubts about the Copenhagen Interpretation. The two became firm friends, bouncing ideas off each other, and in spite of his doubts about the road quantum physics was taking, the next few years were a golden time for Schrödinger.

The golden years

The University of Berlin had been in existence since 1809. It was the brainchild of the explorer and naturalist Alexander von Humboldt (1769–1859), the son of a Prussian officer and statesman, and a member of the higher echelons of Prussian society (he became Baron von Humboldt) as well as a great scientist. With the help of his elder brother, a lawyer and diplomat, he made a proposal to the king, Friedrich Wilhelm III, for the establishment of a new university, which the king endowed with both funds and a home in a former palace on Unter den Linden.

By the second half of the 1920s, the German economy had recovered from its post-war collapse and was enjoying a period of relative calm before the storm that was to engulf the country in the 1930s. Although the country was governed by a variety of short-lived coalitions, from 1925 the President, providing an image of stability, was the popular war leader Field Marshal Hindenburg, a symbol of many Germans' belief in an army that had not, according to their mythology, been defeated on the battlefields of the Great War. The Berlin of the late 1920s was a curious mixture of the artistic and the pornographic, a city where "anything goes"—the city of Bertolt Brecht and Kurt Weil's *Threepenny Opera*. And science flourished there alongside the arts.

At the end of the 1920s, the physics faculty in Berlin was second to none. Planck, although retired, continued to give lectures as emeritus professor; Lise Meitner (1878–1968), who was the first woman in Germany to be appointed as a full professor of physics and played a key part in the discovery of nuclear fission, lectured on nuclear physics; Walther Nernst (1864–1941), a leading thermodynamicist, taught experimental physics; Max von Laue was there; Fritz London (1900–54), who helped develop the quantum theory of the chemical bond, gave lectures on the subject, hot off the press. In this galaxy of teachers, Schrödinger was top dog. And although Einstein had no lecturing duties, he was there to talk things over with.

It was generally agreed by the students that Schrödinger's lectures were the clearest and the best, given without notes from a master at his craft. He was also popular with the students for his informality, both in his style of teaching and in the way he dressed. His informality in the matter of clothing had been commented on at the Solvay Congress (not least by Dirac, who was always "correct"). With the confidence born of his success with wave mechanics, and secure in his appointment in Berlin—not just a job for life but *the* job for life—he felt no need to dress to impress, and turned up at the smart hotel in Brussels where the participants were staying clad for hiking, with his knapsack on his back. In the official conference photograph, among the serried ranks of serious scientists in dark sober suits, Schrödinger stands out (even though he has been carefully placed at the back to hide him as much as possible) in a light casual jacket. The wonder is that he bothered with the jacket at all. In Berlin, which was still steeped in the formal Prussian tradition which prescribed a dark suit and white shirt with collar and tie as de rigueur for lecturers, Schrödinger addressed his students wearing a sweater and a jaunty bow tie in winter, and an open-necked short-sleeved shirt in summer. On one famous occasion, when he failed to turn up on time for a lecture the students sent out a search party to look for him, and found that the security guard had refused to admit the scruffy "imposter"; only with some difficulty was the gatekeeper persuaded that this really was Herr Professor Schrödinger, and that the students were waiting to hear him talk.

The Schrödingers were no less popular with their peers, and threw large parties which became a feature of the academic social scene. But they seem to have been less popular with each other, and Schrödinger's colleagues who visited him at home individually noticed that the couple barely spoke to one another. Erwin had several romantic affairs while in Berlin (he was always romantic, always convinced that he was in love), and also saw Itha Junger whenever possible on his travels outside the city. In 1929, a new possibility appeared on his romantic horizon when he visited Innsbruck to give a lecture and stayed with the recently married physicist Arthur March (1891–1957). Erwin was struck by the charms of March's bride, Hilde. Nothing happened at the time, but years later Hilde would have a significant effect not just on Schrödinger's private life but on his career.

Meanwhile, the relationship with Itha developed to the point where she came to stay with Schrödinger in Berlin in the summer of 1932, while Anny was away. The result was pregnancy, and an abortion. Probably as a result of this, Itha, who would later marry an Englishman, suffered several miscarriages and was never able to have children. After she left Berlin, Erwin met her only once more, in London in 1934. By then, the Schrödingers were based in Oxford, having left Germany because of politics and been welcomed in England because of science.

Schrödinger's position in the scientific world was cemented in February 1929, when he was elected a member of the Prussian Academy of Sciences. This was an honour that conferred membership of the élite, since membership was restricted to thirty-five people in each of two classes, the Physical-Mathematical Class and the Philosophical-Historical Class. In nominating Schrödinger, who became the youngest member of the Academy at the age of forty-two, Planck referred to the way in which "the hitherto somewhat mysterious wave mechanics [of de Broglie] in one stroke was placed upon a firm foundation." The Academy registered another, less welcome, sign of Schrödinger's eminence a few years later, when the Nazis came to power in Germany. The only two names expunged from the membership records of the Academy, as if they had never existed, were those of Einstein and Schrödinger.

The discovery of wave mechanics was, of course, an impossible feat to top, and nobody expected Schrödinger to hit those heights again. But during his time in Berlin Schrödinger's idea received a boost from Dirac, and Schrödinger himself made a suggestion which, although few people at the time took it seriously, was later revived, or rather rediscovered, in a big way. It was connected to the "Dirac Equation."

Dirac's breakthrough came late in 1927. By then the concept of electron spin had been around for several years, and physicists were becoming aware that it was a purely quantum property which had nothing to do with spin, or rotation, in the everyday sense. The big drawback of the Schrödinger wave equation was that it did not include spin. Tweaking the equations of quantum mechanics by dropping spin in by hand helped to make predictions that matched the results of experiments, but nobody had yet been able to come up with an equation that incorporated spin as an integral part of the theory. This was what Dirac now did. He found an equation which described the behaviour of the electron, taking full account of the requirements of relativity theory, and producing the property known as electron spin. This fully relativistic electron equation, now known as the Dirac Equation, did not predict anything new, but it explained everything in one mathematical package with no need for added extras.

Or did it predict something new? The puzzle was that the equation seemed to have two solutions, one positive and one negative, in the same way that a number such as 4 has two square roots—in this case, 2 and –2. The Dirac Equation seemed to be predicting the existence of both everyday electrons and negative electrons—and since an electron has negative electric charge, and two negatives make a positive, a negative electron would have positive charge. Nobody knew quite what to make of this until 1932, when the American Carl Anderson (1905–91) discovered a counterpart to the electron, but with positive charge instead of negative charge, while studying cosmic rays. It was soon dubbed the positron, or anti-electron. We now know that for every kind of particle there is a kind of anti-particle—an anti-proton, an anti-neutron, and so on. But while particles (such as electrons) are common in our world, anti-particles (such as positrons) are rare.

To his contemporaries, the most interesting work Schrödinger did during his time in Berlin stemmed from Dirac's new work. He investigated the properties of the electron as described by the Dirac Equation, and published two papers on the theme, in 1930 and 1931, through the Prussian Academy. His results are too esoteric to go into here; but in March 1931 Schrödinger offered the Academy a remarkable paper which contains a breathtakingly simple, but almost unbelievable, idea, in the same vein as Dirac's negative electrons.

Back to the future

This new speculation stemmed from Schrödinger's early fascination with thermodynamics and the way the world is governed by statistical laws. He noticed that his wave equation had a similar structure to the equation used to describe diffusion processes, such as the way molecules of perfume from an open bottle spread through the air. Thanks to his work in statistical mechanics, he was also aware of a curious property of the diffusion equation. Such an equation can be run in reverse, to describe a world in which molecules of perfume come together out of the air and congregate in an open bottle, even though we do not see events like that going on in the everyday world. This kind of reversibility lies at the heart of the statistical understanding of thermodynamics developed by Boltzmann and others. Now, you might think that if you had all the information about the distribution of scent molecules through the air at a certain time, and the equivalent information for a later time, you could calculate the distribution of the molecules for any time in between either by working forward from the earlier time or backward from the later time. But, as Schrödinger realized, you would be wrong. The way to find the distribution for intermediate times is to combine the solution for the equation going forward in time with the solution to the equation going backwards in time—in effect, multiplying the two equations, or their solutions, together.

The connection with quantum mechanics—what Schrödinger described in his paper as "the most interesting thing about our result"—comes from the way the square of Schrödinger's wave function is used to calculate probabilities in the Copenhagen Interpretation. The wave function is usually denoted by the Greek letter *psi* (ψ), but the square involved in the calculations is not simply $\psi \times \psi$. Because the equations, like all good wave equations, involve the square root of minus one (i), and are therefore "complex" in the mathematical meaning of the term, the wave function has to be multiplied by something known as its complex conjugate, which can be denoted by ψ^*. So the probability of finding an electron at a particular place depends on $\psi \times \psi^*$. But the complex conjugate is, in effect, the same as the wave function running backwards in time. Like the solution to the diffusion problem, the probabilities in the Copenhagen Interpretation depend on combining two equations, one describing processes proceeding forward in time and one describing processes proceeding backwards in time.

At this point Schrödinger ran into a brick wall, and rather lamely concluded: "I cannot foresee whether the analogy will prove useful for the explanation of quantum mechanical concepts." Nor could anyone else in the 1930s, and

Schrödinger's paper stirred so little interest that, half a century later, when the American physicist John Cramer (b. 1934) did find a way to interpret the complex conjugate to provide a new understanding of quantum mechanics he did so in complete ignorance of Schrödinger's 1931 paper.

I have described Cramer's "transactional interpretation" of quantum mechanics fully in my book *Schrödinger's Kittens*, but it is worth going into a little detail here since it shows just how deep Schrödinger's insight was. Although Cramer was unaware of this particular insight, one of the ideas that had set him thinking about waves travelling backwards in time was nearly as old. In 1940, Richard Feynman was a graduate student at Princeton University, working under the supervision of John Wheeler (1911–2008). He became interested in a problem known as radiation resistance, which in simple language means that it is hard to push charged particles like electrons around—they resist, more strongly than uncharged particles resist, and at the same time they radiate electromagnetic waves. But Feynman knew that Maxwell's equations, which describe all electromagnetic radiation, are symmetrical in time (essentially for the same reason that Schrödinger's wave function is symmetrical in time, although he did not make that connection in 1940). He suggested, backing up the suggestion with calculations, that when an electron (or any charged particle) is jiggled about, it radiates electromagnetic waves both into the future and into the past. Wherever, and whenever, this radiation meets another electron (or other charged particle), it makes that particle jiggle about, and spread waves into the past and into the future. The overlapping waves interact with one another and mostly cancel out, like the probability waves in the Copenhagen Interpretation; but some of the waves, from both past and future, return to the original electron and provide the resistance needed to explain the reluctance of charged particles to be pushed around.

Wheeler was sufficiently impressed that he told the twenty-two-year-old Feynman to give a talk explaining his ideas to the physics department at Princeton. This was a daunting task, not least since the audience included both Einstein and Pauli. Pauli was unimpressed, and said that the idea was nothing more than a mathematical tautology; but Einstein replied: "No. The theory seems possible." The work, tidied up with the help of Wheeler, was published in 1941, and became known, though never universally accepted, as the "Wheeler-Feynman" theory of radiation resistance. A mere forty-five years later, John Cramer incorporated these ideas into quantum mechanics.

Most quantum mechanics of the 1980s ignored the physical interpretation of the Schrödinger equation and simply used the probabilities without bothering too much about where they came from. But Cramer went back to the full relativistic version of the equation with two sets of solutions, one corresponding to a wave travelling forward in time and one to a wave running backwards in time. These are known, respectively, as "retarded" waves and "advanced" waves. Within this framework, Cramer described a typical quantum "transaction" (such as an interaction between two electrons) in terms of the particles "shaking hands" with each other across space and time. In order to get a feel for what is going on, you have to stand outside of time, in a sense, and look at the interaction, or transaction, from the perspective of some kind of supertime. There is no suggestion, though, that this supertime is real; it is just a device to help us get the picture straight in our minds.

Imagine a quantum entity that makes a transition from one quantum state to another—an electron, perhaps, which starts in a state corresponding to a position on one side of the experiment with two holes, and ends up in a state corresponding to a point on the detector screen on the other side of the experiment. The electron is described by a wave which spreads out into the future. It is also described by a wave travelling into the past, but ignore that for the moment. The retarded wave arrives at the detector screen, where it triggers the emission of further sets of advanced and retarded waves. The advanced waves from the detector travel back in time to the original position of the electron, where one set of waves is selected at random, in accordance with the rules of probability, and "chosen" to combine with the original retarded wave. This triggers the production of a "new" advanced wave travelling backwards in time which cancels out the original advanced wave (which is why we could ignore it), and a "new" retarded wave travelling into the future, which cancels out the original retarded wave everywhere except where the "handshake" has occurred. So the electron makes a transition from its starting point to one point on the detector screen. But it is meaningless to ask what it is doing "in between"; there is no in between.

As Cramer put it: "The emitter can be considered to produce an 'offer' wave which travels to the absorber. The absorber then returns a 'confirmation' wave to the emitter, and the transaction is completed with a 'handshake' across space-time." But in reality, there is no toing and froing; it all happens instantaneously, producing the effect of what is sometimes known as action at a distance. This "transactional interpretation" of quantum mechanics makes exactly the same (correct) predictions about the outcomes of experiments as the Copenhagen Interpretation, and makes no predictions that differ from those of the Copenhagen Interpretation or other interpretations of quantum mechanics; so ultimately the choice of which one you prefer is a matter of personal taste. But it does show that it is possible to understand quantum mechanics without invoking the "collapse of the wave function"—and that in itself is a thoroughly good thing, since there is nothing in the equations that describes, or requires, such a collapse; it is merely a heuristic device introduced by Bohr on the basis of no evidence whatsoever. And Schrödinger hated it.

As I discuss in the next chapter, Schrödinger's antipathy to the idea of collapsing wave functions led him to his most famous thought experiment. But by the time he came up with the idea he had left Berlin, as a result of the political turmoil that swept through Germany in the early 1930s.

People and politics

The immediate trigger for that turmoil was the global economic crisis that followed the Wall Street crash of October 1929. The recession and resulting mass unemployment in Germany gained the Nazi party many recruits, and led to violent street battles between their supporters and the Communists. Their share of the vote gradually rose during successive elections, peaking at 37 per cent in July 1932, when they became the largest party in the Reichstag, although without an overall majority.

Although tenured university professors were in many ways insulated from these problems, and actually benefited from falling prices as their salaries were maintained, research funding was slashed. Students and junior university staff suffered with the general population, and many reacted the same way. There were riots within university precincts (where the police were not allowed), and demonstrations in support of the idea of a quota limiting the number of Jewish students. Schrödinger made no secret of his dislike of the Nazis, but he was never actively involved in opposing them. He carried on with his teaching, but did little research as things came to a head.

In another election, in November 1932, the Nazis' share of the vote fell back from 37 per cent to 32 per cent. But an unholy alliance of old generals, leaders of the aristocratic families known as the Junkers, and industrialists had decided that the kind of strong leadership and national pride that the party offered would be the right thing for Germany, and pressured Hindenburg, who was still President even though he was now senile and not really aware of what was going on, into appointing Adolf Hitler as Chancellor on 30 January 1933. The die was cast, and if the generals and the Junkers thought that Hitler would be their grateful puppet they were sadly mistaken. In March 1933 Hitler called new elections, and achieved a majority in the Reichstag by the simple expedient of excluding the Communists. He used that majority to pass an act which gave him dictatorial power, and made the assembly redundant.

The troubles now began to have an impact closer to home for Schrödinger. Einstein was in America at the time Hitler was appointed Chancellor, and vowed that he would not return to Germany as long as the Nazis were in power; he also resigned his membership of the Prussian Academy. The

Academy replied, in effect, "good riddance," and Schrödinger, without making any public comment on the situation, stopped attending its meetings. Laws excluding Jews and other "undesirables" from holding government positions, including academic posts at universities, were passed, and within a year nearly two thousand faculty members across Germany were sacked—including Max Born.

Many scientists around the world watched in horror, but felt there was nothing they could do to help. One man in particular, though, decided he must try to assist the expelled Jewish scientists. He was Frederick Alexander Lindemann (1886–1957), then head of the Clarendon Laboratory in Oxford and universally known as "the Prof." Lindemann seems an unlikely figure to leap to the aid of German Jews. He was extremely wealthy, having inherited a fortune from his father, unmarried, and with an unpleasant streak that manifested itself in sarcastic remarks about those he considered his inferiors (including most women). He had no Jewish connections, and like most of his class in England at the time was if anything mildly anti-Semitic. His initial idea seems to have been simply to get hold of some top scientists to enhance the physics faculty at Oxford University. But his scheme grew like Topsy. In spite of the difficult economic situation, Lindemann managed to get funding from the British company Imperial Chemical Industries (ICI) to provide new posts for Jewish scientists with established reputations, so that British scientists would not be adversely affected and no public money would be required. The scheme became a great success, and probably saved many lives; but here I am only interested in the way it affected Schrödinger.

In April 1933, Lindemann visited Germany to assess the situation and draw up a short list of scientists whom he might be able to help. He had thought that the problem would be a temporary one, and that Germany would soon recover from what he called "the Nazi madness"; but what he saw persuaded him that the Nazis would be in power for a long time. During his visit, he met Schrödinger at his home in Berlin, and listened while he spoke of his distaste for the new regime. Lindemann offered one of the new ICI fellowships to Schrödinger's assistant Fritz London, then a *Privatdozent* at the university, but to the surprise of both Lindemann and Schrödinger he asked for time to think things over.[2] So Schrödinger said: "Offer it to me." Lindemann was taken aback. Schrödinger was not Jewish, and at that time under no threat from the regime; but he was so opposed to what was going on in Germany that at the age of forty-five he was willing to give up his job for life and face an uncertain future as an émigré in England on a short-term appointment. It would be a great coup for Oxford, and Lindemann promised to check out the possibilities on his return to England. But there was one catch (although Lindemann did not know it at the time), involving Schrödinger's private life. He asked for additional funding for a post for his friend Arthur March, on the grounds that he and March wanted to write a book together. The real reason was that Schrödinger was by now, with Itha off the scene, in hot pursuit of March's wife, Hilde.

The pursuit continued through the summer, and across Europe, while Lindemann was, among other things, sorting out the funding situation in England. Schrödinger's determination to leave Germany was reinforced in May, when the German government introduced a visa fee of 1,000 marks for

Germans, or state employees such as Schrödinger, wanting to visit Austria. This meant that he and Anny could not afford to visit their homeland, even for the seventieth birthday celebrations of Anny's mother. With hardened resolve, the Schrödingers packed up many of their belongings and sent most to England and some to Switzerland. Soon after making these arrangements, the couple set off on holiday in their new BMW, heading for the Tyrol. They started out with a hired chauffeur/instructor, who taught them to drive as they went along, and left them at the Swiss border. Schrödinger wrote a formal letter to the authorities requesting "study leave," but did not resign his post; he also sent a postcard to the porters at the physics department informing them that he would not be giving his lectures in the autumn. His salary was stopped on 1 September, but by then he was safely out of reach.

So far, Schrödinger's pursuit of Hilde had been unsuccessful, even though he had offered to divorce Anny and marry her. But he was confident. In May, he wrote in his diary: "It has never happened that a woman has slept with me and did not wish, in consequence, to live with me for all her life. I swear in the name of the good God that it will be the same thing with her." And the situation was about to change.

With Anny doing most of the driving, the Schrödingers visited Zürich and went on through the mountains to Italy, to the town of Bressanone, where Arthur March had been born when, under the name of Brixen, it was part of the Austrian Empire. There, they met up with the Marches. The Born family were staying nearby, and Anny went off to visit them with her lover Hermann (Peter) Weyl. Hilde's resistance crumbled, and, apparently with Arthur's acquiescence, she went off on a cycling tour with Erwin. By the time they came back, she was pregnant, having been childless for four years of marriage. This may be why Arthur condoned the arrangement and remained friendly with Erwin; as for Anny, she was hardly in a position to object and remained friendly with Hilde.

It says something about Schrödinger's propensity for flirtation that even while fresh from his successful conquest of Hilde, he was not blind to other opportunities. In September, Erwin moved on to Malcesine, on Lake Garda, having written to Lindemann, who he knew would be visiting Italy, to suggest a meeting there. In Malcesine, he bumped into Hansi Bauer, the daughter of Anny's old employer, now twenty-six and on her honeymoon. She had married Franz Bohm, ten years her senior, who had studied engineering and political science in Vienna and had served, like Schrödinger, as an artillery officer in the First World War. He was now working for the Ingersoll company. According to her account in an interview with Walter Moore, the honeymoon was not going well, and she was already disillusioned by her marriage when she chanced upon Erwin in a grocery shop and there was "a spark" between them. But it would be some time before that spark lit a flame.

Lindemann soon arrived at Lake Garda with good news. He was able to offer Schrödinger an appointment for two years, funded largely by ICI but based in Oxford and carrying out research there, with a package of remuneration roughly equivalent to the salary of a professor. On 3 October, Schrödinger was elected a Fellow of Magdalen College in his absence. Soon afterwards he set out with Anny in the BMW for Paris and Brussels, where he attended the seventh Solvay Congress but took no active part. The couple finally arrived in Oxford on 4 November 1933. The timing was perfect—on the day of his formal welcome as a Fellow of Magdalen, news came that Schrödinger had been awarded the 1933 Nobel Prize in physics, jointly with Paul Dirac. So in more ways than one, it was the beginning of a new phase of his life.

Chapter Nine

The Coming of the Quantum Cat

The Schrödingers barely had time to settle in Oxford before it was time to leave for the Nobel award ceremony; the awards are always presented on 10 December, the anniversary of the death of Alfred Nobel, the inventor of dynamite, who had left his fortune to establish the prizes. They arrived in Stockholm on 8 December, and Erwin duly received his share of the physics prize, 100,000 kroner (about $27,000 at the prevailing exchange rate); wisely, he decided to keep the bulk of the money in Sweden. On 12 December he gave his formal Nobel Lecture, on "The Fundamental Idea of Wave Mechanics," avoiding any deep mathematics, and then it was time for the couple to return to Oxford. They eventually settled in a large rented house at 24 Northmoor Road; the Marches settled nearby, at 86 Victoria Road. But one of the first decisions Erwin had to make was when to take an extended leave of absence from Oxford.

Back in the USA

When he had arrived in Oxford—before learning about the Nobel Prize—Schrödinger had found a letter from Princeton University waiting for him. It contained an invitation to give a series of lectures at Princeton for either one, two, or three months, at his choice. For this he would be paid a salary of $1,000 per month; and there would be a contribution of $500 towards his travel expenses thrown in. In response to an offer like that, roughly equivalent to a year of his Oxford salary for three months' work, the only question was when to leave and how long to go for.

There was an ulterior motive behind such a generous offer. The Physics Department at Princeton was looking for someone to fill the vacant professorship of mathematical physics, and Schrödinger was on a short list of two (himself and Heisenberg). The lecture tour would give them a chance to look at him, and, if they liked what they saw, a chance to persuade him to take up the full-time post. But it turned out that they would have to exercise their powers of persuasion rather swiftly, since Schrödinger took up the offer of a one-month appointment, leaving England on 8 March 1934 and returning on 13 April. In his absence Anny visited the Borns, now in Cambridge.

While in Princeton, Schrödinger stayed at the Graduate College, an imitation of an Oxbridge college housed in a mock-Gothic building. He gave his usual immaculate lectures, and was duly offered the professorship, a post which, ironically in view of the success of his lectures, had no teaching duties. He immediately wrote to Lindemann to inform him of the situation, explaining that he had to consider the offer seriously, "for after all and in spite of all so-called reputation I am actually without what a man of my age and métier considers a [permanent position]. And if, e.g., I were drowned on the passage [home], I am afraid that my wife could neither live upon the German pension nor on the [Schrödinger equation]."

This letter highlights Schrödinger's increasing concern with establishing long-term security, not so much for himself but for his wife after his death, a concern based on what he had seen happen to his own mother. Getting a job with a good pension became almost an obsession with him, and would soon lead him into an unwise move. But even for a pension Schrödinger was reluctant to live in America, and stalled as long as possible before finally turning the offer down in June (Princeton didn't get Heisenberg, either). Significantly, although the salary on offer at Princeton was excellent ($10,000, roughly twice what Schrödinger earned in Oxford), the widow's pension associated with the post was not very good, amounting to only $200 a year.

And there was another snag. What to do about the pregnant Hilde March? In his private journal, Schrödinger wrote that he would be sad to "leave the mother and child," and there was certainly no prospect of arranging a post for Arthur March in Princeton. It is widely believed in Princeton (as a matter of folklore rather than proven fact) that Schrödinger discussed the problem with John Hibben, the Princeton University President, who was horrified at the thought of Schrödinger bringing a wife, a mistress, and an illegitimate child to the conservative campus. The Oxford establishment was scarcely less horrified at what they had let into their midst.

Oxford and beyond

In the male-dominated environment of Oxford colleges in the 1930s, it was considered a bit odd to have one wife, let alone two. But Schrödinger treated Hilde exactly as a second wife, going around with her in Oxford in the months leading up to the birth of their child, and making no secret of their relationship. Their one concession to conventional morality was that when the baby, Ruth George Erica, was born, on 30 May 1934, the name of her father given on the birth certificate was Arthur March. Hilde, suffering perhaps from post-natal depression, or maybe the strain of living in the Schrödinger ménage, or both, had little interest in the baby initially, and she was largely looked after by Anny (and, of course, a nurse) for the first few months.

All of this put a severe strain on Schrödinger's relationship with Lindemann and ICI, who felt (correctly!) that they had been hoodwinked into appointing March. The antipathy between Schrödinger and Oxford was mutual. As he explained to Max Born,[1] "These colleges are academies of homosexuality. What queer types of men they produce." And he was baffled by the informality of college dinners. "You never know who your neighbour might be. You talk to him in your natural manner, and then it turns out that he is an archbishop or a general." Nor was he mollified by his teaching duties—or rather, the lack of them. He only had to give one lecture a week, on "Elementary Wave Mechanics," and grumbled to Anny that there was so little work to do that he felt he was being treated as a charity case. But in the summer of 1934, he had a chance to get away from it all.

Schrödinger was invited to give lectures in Spain, at Santander and Madrid, and seized the opportunity, while

Anny seized her own opportunity to meet up with Weyl in Switzerland. The Santander lectures were translated into Spanish and published as a book; even more significantly, Schrödinger struck up a rapport with the philosopher José Ortega, who had organized the meeting. The whole experience was such a delightful contrast with Oxford that in the spring of 1935 he returned, with Anny and the BMW, to make a proper tour of the country, stopping off to give another series of lectures in Madrid. There seems to have been a real possibility of a permanent move to the University of Madrid, but any such prospects were quashed by the outbreak of the Spanish Civil War in 1936; Schrödinger's friend Ortega, a staunch Republican, was among those eventually forced into exile.

It may have been with the possibility of a move to Spain in mind that Schrödinger tidied up a loose end while he was in Madrid. He wrote to Berlin formally resigning his professorship there. The resignation was formally accepted on 31 March; on 20 June, Hitler sent a formal letter of thanks to Schrödinger for his services, and in July 1935 he was given the honorary title of professor emeritus. All seemed to be sweetness and light between Schrödinger and the authorities in Berlin.

Back in England, Schrödinger was invited to give a talk in a series on freedom, a hot topic with the rise of fascism in Europe, broadcast on BBC radio. This is the sort of thing Nobel laureates are asked to do, whatever their speciality. His title was "Equality and Relativity of Freedom," and the talk was published, along with others in the series, in *The Listener*. His main point, that freedom is relative, hardly seems world-shattering news, but the talk is significant in the context of

Schrödinger's life for one omission and one inclusion. The omission is any mention of the situation in Nazi Germany at the time; the inclusion shows how deeply Schrödinger was affected by Itha's abortion, although (ironically revealing the suppression of another freedom) in order to refer to it in a BBC broadcast of the 1930s he has to use circumlocutions: "The individual, this time a female, in order to avoid contempt and rejection by all the 'respectable people,' is compelled to commit an action, which is threatened by the law of most countries with penal servitude. Most of you will know what I mean."

Just after he made this broadcast, Schrödinger's attention was focused back on the interpretation of quantum mechanics by the latest contribution from Einstein.

Faster than light?

By 1935, Einstein was settled in Princeton, at the Institute for Advanced Study. He had been working with two younger colleagues, Boris Podolsky (1896–1966) and Nathan Rosen (1909–95), and together (led by Podolsky on this occasion) they had come up with what seemed to them an unarguable refutation of the nonsense (as they saw it) inherent in the idea of collapsing wave functions and the Copenhagen Interpretation. Their paper describing what became known as the "EPR Paradox," even though it is not really a paradox, appeared under the title "Can Quantum Mechanical Description of Physical Reality Be Considered Complete?" in the journal *Physical Review* in May 1935.[2] They described the puzzle in terms of measurement of position and momentum, but I shall use what seems to me a simpler example involving electron spin.

Imagine a situation in which two electrons are ejected from a quantum system (such as an atomic nucleus) in different directions, but are required by the laws of symmetry to have opposite spin. According to the Copenhagen Interpretation, neither of the electrons possesses a definite spin until it is measured; each exists in a 50:50 "superposition" of spin up and spin down states, until it is measured. Then, and only then, the wave function collapses into one or the other state. But in this example the laws of symmetry require the other electron to have the opposite spin. This is fine when both electrons are in the superposition of states, but it means that at the instant one electron is measured, the other electron, which might by now be far away (in principle, on the other side of the Universe), collapses into the opposite state at the same instant. How does it know to do this? It seems that what Einstein called a "spooky action at a distance" links the two particles, which communicate with one another faster than light. And all quantum entities (which means everything) must be linked in the same way.

It is a key tenet of the theory of relativity, which has passed every test ever applied to it, that no signal can travel faster than light; so Einstein, in particular, saw this argument as a complete refutation of Bohr's ideas. The EPR paper concluded that the Copenhagen Interpretation makes the reality of properties of the second system "depend upon the process of measurement carried out on the first system, which does not disturb the second system in any way. No reasonable definition of reality could be expected to permit this."

The alternative that Einstein favoured is that there is some kind of underlying reality, an invisible clockwork which controls the workings of the Universe and gives the appearance of uncertainty, collapsing wave functions, and so on, even though "in reality" each of the electrons, in this example, always has a well-defined spin. In other words, things are "real," not in a superposition of states, even when we are not looking at them. The idea that the Universe is composed, even at the quantum level, of real things that exist whether or not we observe them, and that no communication can travel faster than light, is known as "local reality."

It is, perhaps, just as well Einstein did not live to see a series of beautiful experiments carried out in the 1980s which proved that local reality is not a good description of the Universe. I'll go into these in more detail later, but their implication is that we are forced to abandon either the local bit (allowing communication faster than light) or reality (invoking instead collapsing wave functions)—or, as I shall explain, go for something else entirely. But nobody knew this in 1935, and Schrödinger in particular was delighted when he saw the EPR paper. He wrote at once to Einstein, commenting that "my interpretation is that we do not have a q.m. that is consistent with relativity theory, i.e., with a finite transmission speed of all influences," and in a paper published in the *Proceedings of the Cambridge Philosophical Society* later that year said: "It is rather discomforting that the theory should allow a system to be steered or piloted into one or the other type of state at the experimenter's mercy in spite of his having no access to it."[3] This was the genesis of Schrödinger's famous cat.

The cat in the box

The ideas encapsulated in the famous "thought experiment" involving Schrödinger's cat actually came in no small measure from Einstein, in the extended correspondence between the two triggered by the EPR paper and preserved in the Einstein Archive at Princeton University. Einstein introduced the idea of two closed boxes and a single ball, "which can be found in one or the other of the two boxes when an observation is made" by looking inside the box. Common sense says that the ball is always in one of the boxes but not the other; the Copenhagen Interpretation says that before either box is opened a 50:50 wave function fills both of the boxes (but not the space in between!), and when one of the boxes is opened the wave function collapses so that now the ball is in one box or the other. Einstein continued: "I bring in the separation principle. The second box is independent of anything that happens to the first box."

In a later letter, Einstein came up with another *reductio ad absurdum*. He suggested to Schrödinger the idea of a heap of gunpowder that would "probably" explode some time in the course of a year. During that year, the wave function of the gunpowder would consist of a mixture of states, a superposition of the wave function for unexploded gunpowder and the wave function for exploded gunpowder:

> In the beginning the ψ-function characterises a reasonably well-defined macroscopic state. But, according to your equation, after the course of a year this is no longer the case at all. Rather, the ψ-function then describes a sort of blend of not-yet and of already-exploded systems. Through no art of interpretation can this ψ-function be turned into an adequate description of a real state of affairs . . . in reality there is just no intermediary between exploded and not-exploded.

Stimulated by the EPR paper and his correspondence with Einstein, Schrödinger wrote a long paper, published in three parts in the journal *Die Naturwissenschaften* later in 1935, summing up his understanding of the theory he had helped to invent. It was titled "The Present Situation in Quantum Mechanics," and it introduced to the world both the term "entanglement" and the cat "paradox," which (like the EPR "paradox") is not really a paradox at all. An excellent English translation of the paper, by John Trimmer, appeared in the *Proceedings of the American Philosophical Society* in 1980, and can also be found in the volume edited by Wheeler and Zurek, *Quantum Theory and Measurement*. Many garbled accounts of the "cat in the box" "experiment" have appeared over the years, but it is best to go back to this source and Schrödinger's own words (as interpreted by Trimmer) to get the puzzle clear:

> One can even set up quite ridiculous cases. A cat is penned up in a steel chamber, along with the following diabolical device (which must be secured against direct interference by the cat): in a Geiger counter there is a tiny bit of radioactive substance, so small, that perhaps in the course of one hour one of the atoms decays, but also, with equal probability, perhaps none; if it happens, the counter tube discharges and through a relay releases a hammer which shatters a small flask of hydrocyanic acid. If one has left this entire system to itself for an hour, one would say that the cat still lives if meanwhile no atom has decayed. The first atomic decay would have poisoned it. The ψ-function of the entire system would express this by having in it the living and the dead cat (pardon the expression) mixed or smeared out in equal parts.

> It is typical of these cases that an indeterminacy originally restricted to the atomic domain becomes transformed into macroscopic indeterminacy, which can then be resolved by direct observation.

In other words, according to the version of quantum mechanics that was generally taught and widely (but not universally) accepted for the rest of the twentieth century, the cat is both dead and alive (or if you prefer, neither dead nor alive) until somebody looks inside the chamber and by the act of observation "collapses the wave function." But there is nothing in the equations about collapsing wave functions. Remember, this is an entirely ad hoc idea introduced by Bohr, with no basis in reality. That is the single most important message to take away from Schrödinger's thought experiment (which, I stress, is indeed "all in the mind"; nobody has ever done anything like this to a real cat). Although the "cat in the box" idea did not generate widespread interest in 1935, Einstein at least fully appreciated the importance of Schrödinger's puzzle; Schrödinger described the idea to him in a letter, before his paper was published, and Einstein replied: "Your cat shows that we are in complete agreement concerning our assessment of the character of the current theory. A ψ-function that contains the living as well as the dead cat just cannot be taken as a description of a real state of affairs."

Schrödinger was right to point out the nonsensical nature of the concept of the collapse of the wave function, and there are much better ways to understand the workings of the quantum world—the most intriguing of which Schrödinger himself later came close to developing. But in the months following the publication of his three-part paper, Erwin had other things on his mind.

From Oxford with love

ICI had originally offered funding for the refugee scientists for two years only, as an emergency measure, on the assumption that this would give them time to find permanent posts. Some did, but many did not. When their two years ran out, Arthur March, Hilde, and baby Ruth had to return to Innsbruck, where Hilde stayed in a sanatorium for several months, recovering from the strain of being Schrödinger's second "wife" and the mother of his child in disapproving Oxford. Schrödinger soon found consolation. Franz Bohm and Hansi had been living in Berlin, but they came from a wealthy Jewish background, and had got out from under the Nazi threat to live in London. This was particularly convenient, since Anny Schrödinger had obtained the use of a flat in London, where she could go to get away from Erwin and leave him more time with Hilde; so he now had the freedom to "be with" Hansi, who became a frequent visitor to Oxford. In the summer of 1935, Hansi and Erwin even went on holiday together, to the Channel Islands.

Professionally, though, things were less satisfactory. Although ICI had been persuaded to make Schrödinger a special case, and offered him two more years' funding, the problem of a pension loomed large, and he had never felt at home among Oxford's collegiate community. The best offer he had received was of a professorship in Madrid, and he would probably have accepted it if the Spanish Civil War had not broken out in July 1936. But there was another prospect on the back burner.

A vacancy was known to be coming up at the University of Graz, in Austria—Schrödinger's homeland. He was seriously tempted, provided that he could be allowed to have a few practical privileges, in particular keeping his money in Sweden. In May 1935, he had written to Einstein:

> It is not that I can't stand it in one place for long. Up till now I've generally been contented wherever I was except in Nazi Germany. Also it is not that they haven't been very nice and friendly to me here. But nonetheless the feeling grows stronger of having no employment and living on the generosity of others. When I came here I thought I would be able to do something for the teaching, but no value is placed on that here . . . I am sitting here waiting for the demise or the complete decrepitude of a very dear old gentleman [the incumbent in Graz] and the possibility that they might make me his successor.

In December, Schrödinger made a visit to Austria, partly to test the waters. He stayed in Graz, gave lectures in Vienna, and visited the Ministry of Education to talk about the possibilities and his special requirements. Over the Christmas period, he combined skiing with a chance to see Hilde and Ruth, and then managed to squeeze in a visit to Max von Laue in Berlin before heading back to England. But there, his developing plans for a return to Austria were temporarily thrown into confusion by another offer.

Charles Darwin, an eminent physicist and grandson of "the" Charles Darwin, resigned his professorship at the University of Edinburgh towards the end of 1935, to become the Master of Christ's College, Cambridge. A committee was set up to seek his successor, and in May 1936 it reported to the University Court that Schrödinger was the best available candidate. As part of the process of finding the right man for the job, the committee had invited Schrödinger to visit

Edinburgh, where he was shown around by Darwin, who had not yet left to take up his new position. Some of the senior members of the university raised their proverbial eyebrows at Schrödinger's clothing: he came, as was his habit, dressed for a hiking tour of the Alps—not the conventional attire for what amounted to a job interview at a British university. But they liked what they saw sufficiently for a formal offer to be made, subject to a guarantee from the Home Office that Schrödinger would be allowed to take up permanent residence in Britain. The salary offered was a modest £1,200 a year; but he would not have to retire before reaching the age of seventy, and there was a good pension attached to the post.

Schrödinger was initially enthusiastic about the prospect of a move to Scotland, and he told Hansi that he would accept the offer if she would come with him. But he must have known there was no real chance of her accepting his invitation, since by now she was pregnant with her first child (his or Franz's? Nobody knows) and about to return to Vienna. Perhaps that was a factor in now making Graz look more attractive than Edinburgh. Whatever the reasons, Schrödinger cooled towards the Edinburgh post, while the Home Office took an interminable time to make up its mind. Before it did so, a formal offer came from Austria—the professorship at Graz, combined with an honorary (but paid) professorship at the University of Vienna. Tactfully, Schrödinger told Edinburgh that the chance to return to his homeland was too good to miss; he did not mention that the chance to be near Hilde, baby Ruth, and Hansi would also be too good to miss. But Edinburgh, although effectively jilted at the altar, didn't do too badly: Darwin's professorship went to Max Born, who stayed with the university (with the Home

Office's approval) in a mutually satisfactory relationship until he retired in 1953. Schrödinger must have left England in the summer of 1936, ready to take up his teaching duties in Graz on 1 October, with a happy heart. But he would return, much less happy, exactly two years later.

CHAPTER TEN

There, and Back Again

With hindsight, Schrödinger's decision to return to Austria in 1936 was, to say the least, ill-judged.[1] Even without the benefit of hindsight, Schrödinger should have known what he was letting himself in for, and he later described his action as "an unprecedented stupidity."[2] Of course he wanted to return to his homeland, and he wanted the security of a pension; but even by the standards of Austria in 1936, Graz in general and the university especially were hotbeds of Nazism. The Professor of Physical Chemistry there was also the leader of the local Nazi party; more than half of the students were active in the party; and the local newspapers were hardline supporters of Nazism. Anti-Semitism was rife, and many Austrians welcomed the attentions of Adolf Hitler, himself an Austrian by birth, who made no secret of his desire to absorb Austria into a greater Germany. All too many people saw this as the closest thing to a rebirth of the Austrian Empire that they could hope for.

From 1932, the Chancellor of Austria had been Engelbert Dollfuss, of the Christian Socialist party. As in "National Socialist," the formal name of the Nazi party, the term "socialist" in the name is somewhat misleading: the Christian Socialists were violently opposed to the main opposition party, the left-wing Social Democrats, but also to the Austrian Nazis, perceived as a threat to their own hold on power. In March 1933 Dollfuss suspended parliament, and Austria became a fascist state more or less along the lines of Italy under Mussolini. At that time, Mussolini, as the first of the fascist dictators, still had a great deal of influence, and supported Dollfuss against both the Nazis and the real socialists. Britain and France, however, were committed to appeasement, and advised Dollfuss not to pick a fight with Hitler. Instead, with Mussolini's encouragement, the Austrian army and police cracked down on the socialists in February 1934, killing many hundreds of activists and imprisoning many others. With the threat from the left neutralized, the Austrian Nazis stepped into the vacuum, assassinating Dollfuss on 25 July 1934 and attempting a coup, which failed because Hitler was still too weak to intervene in the face of Mussolini's threatened opposition. The new Chancellor, Kurt Schuschnigg, ruled thanks only to Mussolini's support; as Hitler's power increased and Italy became entangled in its foolhardy conquest of Abyssinia, Schuschnigg's position weakened. By the time the Schrödingers arrived in Graz in 1936, the writing was on the wall.

Whistling in the dark

It wasn't just the Schrödingers who moved to Graz. Having settled in a large rented house while Erwin took up his lecturing duties in October, early in 1937 they were joined by Hilde and Ruth, who were given the third floor of the house as their own. Arthur March was left behind in Innsbruck, and Anny, who had turned forty on New Year's Eve, now stayed with her mother in Vienna much of the time. Anny, though, seems to have had a much more "maternal" and loving relationship with Ruth than the baby's biological mother did.

Schrödinger's inaugural lecture as professor was not the tour de force that might have been anticipated, but basically a reprise of his Nobel lecture, the sort of thing that might be expected from a Grand Old Man of science settling into semi-retirement. The Grand Old Man status was confirmed by a new honour, when Schrödinger was inaugurated into the new Pontifical Academy of Sciences, as a founder member, at a ceremony in the Vatican on 1 January 1937. The other physicists honoured alongside Schrödinger included Bohr, Debye, Millikan, Planck, and Rutherford—not exactly the "Young Turks" of physics at the time.

Schrödinger's research in Graz was also the kind of thing that Grand Old Men with established reputations, tenured posts, and guaranteed pensions indulge in. He became fascinated by the cosmological ideas of Arthur Eddington (1882–1944), a British Grand Old Man whose illustrious career had included explaining the general theory of relativity to the English-speaking world and testing Einstein's theory by making observations of the stars during a solar eclipse in 1919. He was a great popularizer of science, and intrigued by the puzzle of how to reconcile the general theory with quantum mechanics. But by the 1930s he was in his scientific dotage. He had recently come up with a complicated hypothesis claiming to link cosmology with quantum theory and containing as a crucial ingredient (in what he called the "fundamental relation") a calculation of the number of particles in the Universe, *N*. Nobody has ever understood quite how he arrived at this. Surprisingly, Schrödinger took the idea seriously, at least for a time, and gave a talk about Eddington's "theory" to a conference in Bologna in October 1937. In a long correspondence with Eddington, Schrödinger never managed to fathom why the number *N* actually appears in Eddington's fundamental relation as the square root of *N*, and his attempts to find a quantum theory of the Universe over the next few years were essentially a waste of time.

Teaching was more productive. As well as his lectures in Graz, Schrödinger gave lectures and seminars in Vienna one day a week in termtime, and had an apartment there so that he could stay overnight. This gave him an opportunity to meet up with friends, and to discuss both physics and the political situation. One of those friends, Hermann Mark, later described these get-togethers in an interview with Walter Moore. He explained that Schrödinger, whom he regarded then as a socialist, strongly disliked the political situation in Graz, and spent as much time as he could in Vienna, where the Nazis had less influence. Schuschnigg they regarded as a "moderate" but "no match for Hitler."

In the summer of 1937, with no teaching duties, Schrödinger spent more time in Vienna, joining swimming parties on the Danube and going to lively parties. As Mark told Moore, "those were happy days—but most people did not know they were dancing on a volcano." Or maybe it was that they did not want to know; either way, the parallels with the last days of the Austrian Empire are clear. Hilde usually stayed in Graz while Erwin was in Vienna—but Hansi was there, adding to Schrödinger's enjoyment of the city and its surroundings. The enjoyment was to be short-lived.

Reality bites

Hitler's first move came in February 1938, when he demanded that Schuschnigg visit him in the "Eagle's Lair" at Berchtesgaden. In these intimidating surroundings Schuschnigg was forced to agree to handing over control of the Austrian police and foreign policy to the Nazis, in exchange for a promise (which proved as reliable as all Hitler's promises) that Germany would not invade Austria. When he got home, however, Schuschnigg made a defiantly anti-Nazi speech in parliament, which triggered pro-Nazi riots in Graz. As a last throw of the dice, Schuschnigg then called, on 6 March, for a referendum on Austrian independence, to be held just a week later. Hitler's response was to order his army to invade Austria on 12 March. When Schuschnigg turned for help to Britain, he was told by Lord Halifax, the Foreign Secretary, that none would be forthcoming. So on 11 March Schuschnigg resigned, and to avoid bloodshed told his people not to resist the invasion, or *Anschluss*, that duly took place the following day, without a shot being fired. On 14 March Hitler himself was welcomed in Vienna by delighted crowds lining the streets.

What followed was a brutal assault on Jews and intellectuals worse than anything that had happened under the Nazi regime in Germany up to then. Apart from the physical assaults on Jews and the looting of their businesses, in the first few days of the takeover there were some 76,000 arrests in Vienna, and about 6,000 people were fired from civil service and teaching positions at all levels. While this was going on, the head of the Catholic Church in Austria, Cardinal Innitzer, ordered his churches to fly swastika flags and ring their bells in celebration. He was not alone—the Lutherans held services of thanksgiving for the *Anschluss*.

In the Nazi stronghold of Graz things were calmer than in Vienna, because the *Anschluss* merely consolidated the status quo. But the university was closed; its Rector and dozens of faculty members were dismissed, with some of the Jews being sent to prison and a few managing to escape to other countries, leaving all their possessions and money behind. At first, Schrödinger did not seem to be directly threatened, but with travel restrictions in place he was forced to abandon plans for a visit to Oxford in the autumn of 1938. When his English colleagues heard this they feared the worst; early in April George Gordon, the President of Magdalen (where Schrödinger was still a Fellow), initiated enquiries about Schrödinger's welfare through Halifax and the British ambassador in Berlin. But while his friends were trying to help, Schrödinger was muddying the waters.

If he wanted to stay in Austria, Schrödinger would have to play by the new rules. The Nazis had appointed one of their own, Hans Reichelt, as Rector of the re-opened University of Graz, and his first task was to decide which of the remaining members of faculty could stay and which should be "cleansed." Schrödinger's abrupt and (to the Nazis) insulting departure from Berlin had not been forgotten, and Reichelt advised Schrödinger that he should write a penitent letter to the University Senate spelling out his change of heart. He did so, and the Nazis made sure that the letter was published in full in both German and Austrian newspapers on 30 March.

This was a considerable propaganda coup for them, and it is worth quoting the letter in full:

> In the midst of the exultant joy which is pervading our country, there also stand today those who indeed partake fully of this joy, but not without deep shame, because until the end they had not understood the right course. Thankfully we hear the true German word of peace: the hand to everyone willing, you wish to gladly clasp the generously outstretched hand while you pledge that you will be very happy, if in true cooperation and in accord with the will of the Führer you may be allowed to support the decision of his now united people with all your strength.
>
> It really goes without saying, that for an old Austrian who loves his homeland, no other standpoint can come into question; that—to express it quite crudely—every "no" in the ballot box is equivalent to a national suicide.
>
> There ought no longer—we ask all to agree—to be as before in this land victors and vanquished, but a united people, that puts forth its entire undivided strength for the common goal of all Germans.
>
> Well meaning friends, who overestimate the importance of my person, consider it right that the repentant confession that I made to them should be made public: I also belong to those who grasp the outstretched hand of peace, because, at my writing desk, I had misjudged up to the last the true will and the true destiny of my country. I make this confession willingly and joyfully. I believe it is spoken from the hearts of many, and I hope thereby to serve my homeland.

The reference to a ballot box concerned a referendum to be held on 10 April. The vote that day (supervised by the Nazis) was 99.73 per cent in favour of the *Anschluss*; just 11,929 people had the courage to vote "no." Schrödinger now avoided seeing Hansi, who came from a Jewish family, and asked her to burn the love letters he had sent her.

When reports of Schrödinger's letter reached England, his friends assumed that it must have been written literally at the point of a gun, or under worse duress. Gordon was astonished when one of the Fellows of Magdalen returned from a skiing holiday in the Tyrol with news that he had met the Schrödingers there, enjoying a similar spring break, and had had a long conversation with Erwin. He reported that Schrödinger was quite happy to make his peace with the Nazi regime and saw no need to flee the country, though he expressed a strong dislike of the anti-Jewish policy. He would nevertheless be grateful to meet anyone from Oxford who was in Austria, and said that anything that could be done to stress his international position in science would be helpful. Perhaps most surprisingly, Schrödinger seems to have hinted that he hoped for promotion to an important professorship in Vienna, made vacant by the dismissal of a Jewish incumbent. He seems not to have realized that all such posts would now be filled by active Nazis. At least his naïveté did not extend to financial matters, and he took care to avoid having his Swedish funds transferred to Austria.

The facade of normality was maintained throughout most of April 1938, and on the twenty-third Schrödinger attended a celebratory meeting held in Berlin to mark the eightieth birthday of Max Planck. But on his return to Graz he found that on the very day he had been celebrating Planck's birthday with his colleagues in Berlin he had been dismissed from his honorary post in Vienna. He still had his professorship in Graz, which was rapidly becoming a Nazi university devoted to "relevant" courses in things like the application of chemistry for the war machine and training of the SS medical corps. But as a direct result of Schrödinger's dismissal from his Viennese post wheels started turning to his future advantage. Éamon de Valera, Prime Minister of Ireland, had a passion for mathematics and a pet project to establish an Institute for Advanced Studies in Dublin. When he heard of Schrödinger's situation, and realized that he might soon have to leave Austria, de Valera decided to try to make contact with him through intermediaries to offer him a post in the Irish capital.

Early in May, the German foreign minister, Joachim von Ribbentrop, had confirmed to the British ambassador in Berlin that Schrödinger would not be allowed to travel to Oxford, since this might offer him an opportunity of "resuming his anti-German activities." Oxford, for its part, was no longer sure that it wanted Schrödinger, in the light of his "confession" letter. But de Valera had managed to get a message to Anny's mother in Vienna, via a chain of intermediaries, including Max Born and an old friend of the Schrödingers, Richard Bär, based in Switzerland, offering the Schrödingers what amounted to sanctuary in Dublin. On the pretext of making a casual visit to meet up with the Bärs, Anny travelled to Konstanz, on the border between Germany and Switzerland, and passed on their reply. The Schrödingers would come to Dublin, but nobody must be told anything until they had got out of Austria.

Still Erwin didn't act. Blind to the way events were unfolding, he took a summer holiday with Hilde in the Dolomites, and it was only after he returned to Graz, at the end of August, that he was forced into action. First, he was dismissed from his post in Graz. Even then, it was only when he went to Vienna to discuss alternatives with someone identified simply as "a high official" and was casually told, "Well, they won't let you go to a foreign country," that the penny dropped. Within three days the Schrödingers—still, through some oversight, in possession of their passports—were packed and ready to leave. All valuables, including money and Schrödinger's Nobel medal, had to be left behind; so on 14 September they caught the train to Rome with their clothes in three suitcases and ten marks in Erwin's pocket.

The unhappy return

Anny's account of the Schrödingers' flight from the Nazis is preserved in the Dublin archive; after I wrote *In Search of Schrödinger's Cat*, I heard another account from William McCrea, then (in the mid-1980s) a professor at the University of Sussex, but from 1936 to 1944 Professor of Mathematics at Queen's University of Belfast, and a frequent visitor to Dublin.[3] When the Schrödingers arrived in Rome, Erwin had to ask the taxi driver to tip the porter who carried their bags from the train, then got the commissionaire at the hotel to pay the taxi off, having convinced him that he really was a famous scientist and a friend of the well-known Italian physicist Enrico Fermi (1901–54). He then announced at the reception desk that Professor Fermi would pay their bill. The story, said McCrea, rings true, and "is entirely in character" for Schrödinger. Being a member of the Pontifical Academy clearly had its benefits, even if Erwin's splendid chain of office had had to be left in Graz with his Nobel medal. Fermi, summoned by telephone, came to the hotel and gave them some money, but warned that they were scarcely out of danger in Fascist Italy (before the end of the year he would be forced to flee himself) and that letters were likely to be censored.

There was, however, a loophole. From the premises of the Pontifical Academy, inside the Vatican, Schrödinger was able to write to Lindemann, to his friend Bär in Zürich, and to de Valera, informing them that he was in Rome. Letters from the Vatican, recognized by Mussolini as a sovereign state, escaped the attentions of the Italian authorities. It was easy to contact de Valera, since at the time he was President of the League of Nations, and in Geneva on League business. A couple of days later, while at the Academy, Schrödinger received a telephone call from the Irish Embassy, advising them to get out of Italy as soon as possible. Erwin spoke to de Valera himself (for the first time) that afternoon; the political situation was hotting up, with Germany's takeover of the Czech Sudetenland a real threat, and de Valera urged Schrödinger to get to England or Ireland before the likely outbreak of war.

The Irish Embassy provided the Schrödingers with first-class train tickets to Geneva, but as it was illegal to take currency out of Italy at that time they left with only a pound in money. This led to an unanticipated complication. At the border, the train was stopped and the Schrödingers separated from one another and interrogated while their luggage was searched; Anny described it as "the fright of my life." But the problem was not, this time, political. The Schrödingers' passports had visas for travel right across Europe, and the customs authorities could not believe that people with first-class tickets and Europe-wide visas were travelling with only a single pound in currency. Their not illogical conclusion was that the couple must be smuggling valuables in their luggage. When nothing was found, they were allowed to leave, on the same train, which had been held during the search.

De Valera met the Schrödingers in Geneva, where they stayed for just three days before moving on to England through France. The immediate political crisis had blown over, with Britain and France accepting German demands on Czech territory in the notorious Munich Agreement, supposed to bring "peace in our time." But it would take de Valera many months of domestic politicking, alongside his other political activities, to get the Institute for Advanced Studies established in Dublin, and meanwhile the Schrödingers were once again homeless. At the beginning of October 1938 they turned up in Oxford, where it was made quite clear to them how badly Erwin had damaged his relationships there with his confession letter. Even Max Born, now based in Edinburgh and writing to a colleague in Oxford, commented: "How are you supposed to believe a man who has published that pretty letter?"[4] We will never know if there was anything more to the "pretty letter" than naïveté, but it is consistent with the image we have of a man whose primary aim in life was to provide security for himself and his family.

The Schrödingers spent a couple of uncomfortable months in Oxford, staying with friends; Erwin also visited Hansi, who had escaped from Austria, in London. It was clear they were not welcome and there was no prospect of even a temporary post there. But help came from an unexpected quarter. In the middle of November, Erwin visited Dublin to discuss the arrangements for the new Institute and his role in it, and on his return to Oxford he was relieved to find a letter from Belgium offering him a post as visiting professor at the University of Ghent for the academic year that had just started. He took up the offer, and arrived, with Anny, in mid-December 1938.

Belgian interlude

As well as giving lectures at the university, during his time in Belgium Schrödinger attracted visitors from other universities to discuss physics, and travelled to Brussels, Louvain, and Liège to give talks. The most significant new contact he made was with Georges Lemaître (1894–1966) in Louvain. Lemaître was a pioneering cosmologist who also happened to be an ordained priest, and is often referred to as the "father of the Big Bang," since he was an early proponent of the idea that the Universe as we know it has expanded from a hot, dense state. First hints of the expansion, the famous cosmological redshift which reveals how rapidly galaxies are receding from one another, had been discovered just over ten years earlier, and in a paper in *Nature* in 1939 Schrödinger added his weight to the discussion of what the redshift meant, concluding that it must indeed be caused by the expansion of the Universe.[5] Having become interested in cosmology, Schrödinger made one of the first attempts to integrate cosmology and quantum physics in one package. He did not achieve any notable success, but in a paper published in the journal *Physica* in October 1939 he made a reference to "the production or annihilation of matter, merely by expansion" of the Universe. The effect he discovered does not apply if the Universe is expanding at a steady rate, but is important if the expansion is accelerating. It is intriguing that the present "best buy" idea about the very early Universe is that it went through a phase of rapidly accelerating expansion, called inflation, during which all of the mass-energy was produced; and it has recently been discovered that after a long interval of steady expansion the Universe is now starting to accelerate once again.

Things were also going well, if only briefly, at a personal level. Arthur March brought Hilde and Ruth to Belgium, leaving them there when he returned to Innsbruck. And the University of Ghent awarded Schrödinger his first honorary degree. But all of this was taking place against a background of increasing political tension, and Schrödinger and his extended family were still in Belgium when Germany invaded Poland on 1 September 1939, prompting Britain and France to declare war two days later. “Schrödinger,” said McCrea, “had a singular talent for taking risks.” Although Schrödinger’s time in Ghent was coming to an end, as far as the British were concerned he was now an “enemy alien,” and strings had to be pulled by both Lindemann (in spite of his anger at the confession letter and distaste for Schrödinger’s private life) and de Valera before the whole of Schrödinger’s “family” were granted 24-hour visas to travel through Britain to Dublin, where they arrived on 6 October 1939.

Chapter Eleven

"The Happiest Years of My Life"

Schrödinger spent the next seventeen years based in Dublin—the longest time he had spent living in the same city since his Viennese childhood. He later described these years as "the happiest of my life." He made no more major contributions to physics—hardly surprising, since he was already fifty-two in 1939—but he did make a surprisingly significant contribution to the development of biology, enhanced his reputation as a lecturer, enjoyed a full private life, and made the Institute for Advanced Studies at least temporarily a major centre of world physics. But neither he nor the Institute would have been there if it had not been for the towering figure of Irish politics in the first half of the twentieth century, Éamon de Valera.

"Dev"

De Valera, who was universally known as "Dev," had been born in New York in 1882, the son of an Irish mother and a

Spanish father. His father died when the boy was three, and he was taken to Ireland, where he was raised as a devout Catholic by his maternal grandmother, in a cottage in County Limerick. He studied mathematics in Dublin, became a passionate enthusiast for the Irish language, and married Sinéad Flanagan, who had taught him Gaelic. But instead of becoming an academic, Dev became involved in the active opposition to British rule, and in 1916 he took part in the Easter Rising in Dublin. This militarily futile and violent "rebellion" was not supported by the majority of the Irish people, and resulted in the destruction of the centre of the city. But the brutal response of the British military authorities, who executed many of the leaders of the rising, created martyrs and a groundswell of anti-British feeling. De Valera was among those sentenced to death; but before the sentence could be carried out orders came from an alarmed government in London to halt the executions. Had he not been the last of the leaders of the uprising to surrender, de Valera would have been shot before the new orders from London arrived. Instead, he was imprisoned in Britain, but released during an amnesty when the United States entered the First World War; part of his sentence was served in Lewes gaol, where he passed the time by writing an original mathematical paper, although this was never published.

Continuing his political activities, now as leader of the republican party Sinn Féin, de Valera was arrested and imprisoned once again, but escaped and took part in the Irish civil war of the early 1920s that followed the Anglo-Irish Treaty of 1921, when those seeking full independence clashed with those seeking only home rule within the British Empire (Ireland did have its own parliament, but with limited powers). Like other civil wars, this one set brother against brother and produced bitterness which persists to the present day; de Valera's republicans lost and he spent another spell in prison. After his release in 1924 he founded a new party, Fianna Fáil, dedicated to establishing an independent Irish republic through politics rather than violence, and was elected to the Dublin parliament, the Dáil, in 1927. In 1932, Fianna Fáil and its Labour partner formed a coalition government in Dublin, and by 1937 de Valera, as Taoiseach (Prime Minister), was able to establish the Irish Republic as an independent entity, although still with some links to the British Commonwealth. Those links were not finally severed until 1948, which, as we shall see, led to a curious delay in Schrödinger's being honoured by election to the Royal Society.

Even with all of this going on, Dev nursed two dreams. One was to establish a world-class centre for theoretical physics in Ireland; the other, to revitalize the Irish language. In 1930, when the Institute of Advanced Studies was established in Princeton, not least to provide a home for Albert Einstein, de Valera took note. When he was at last in a position to do something about it, he looked into the possibility of founding such an institute in Dublin, initially with two schools, Celtic studies and theoretical physics. The key to establishing the new institute as a centre of excellence would be attracting a top physicist—someone as near as possible in status to Einstein. Which is why, when Dev heard of Schrödinger's problems in Austria, he acted immediately.

By 1938, Fianna Fáil had a clear majority in the Dáil, and Dev was in a position to push through legislation for what many people regarded as something of a vanity project. The appropriate bill was put before the Dáil on 6 July 1939, spelling out that "the schools will be devoted solely to the advance of learning . . . which will bring students of the postgraduate type from abroad." In a speech to the Senate, the Taoiseach said that although this might seem an inappropriate time for such a bill, it should be regarded as a gesture "to indicate that there is a better way than war for advancing the welfare of mankind." But even Dev was shaken by the wind of change blowing across Europe; with other matters taking priority the bill did not become law until 19 June 1940, and Schrödinger, the first Professor in the School of Theoretical Physics, was not able to take up his post until that October. This gave him almost a year to get settled in Dublin, making contacts and becoming a well-known figure there.

Settling in

Erwin, his two "wives," and five-year-old Ruth settled in the Dublin suburb of Clontarf, near the sea, in a typical middle-class semi-detached house with bay windows. Until 1943, the house was rented; but then Erwin was able to buy it (for £1,000), and he sold it (for £2,150) when the family left Dublin in 1956. Anny and Hilde took turns, one week on and one week off, at doing the domestic chores.

You might expect Erwin's unusual domestic arrangements to have been even more of a problem in Catholic Ireland than in Oxford; but in Dublin at least there was a marked contrast between what was officially approved and what people actually did. The situation is memorably summed up in the words of an Irish Catholic friend of mine, who referred to "sowing her wild oats on weekdays and praying for a crop failure on Sunday." According to McCrea, "Schrödinger and his household were apparently made to feel at home in Clontarf" in spite of his "irregular" arrangements. And it was entirely in keeping with the Irish attitude of *laissez-faire* that one of the first friends Erwin made in Dublin was Monsignor Paddy Browne, a priest who taught mathematics to would-be priests at St. Patrick's College. Msgr. Browne's brother was a cardinal, but that didn't stop Paddy becoming Erwin's best friend in Ireland. Life in Dublin during what was known in Ireland as "the Emergency" was safe and comfortable, apart from a shortage of tea throughout the war, and a growing shortage of coal and oil, which led to private cars disappearing from the streets after 1942. Since Schrödinger was a keen cyclist and hiker, this never bothered him.

Schrödinger began lecturing in Dublin in November 1939, on an informal basis at University College. This was the first opportunity for Dubliners to learn about quantum mechanics from one of its founders, and the lectures were packed. He also continued his research into aspects of quantum theory, and began to publish a series of technical papers in the *Proceedings of the Royal Irish Academy* (RIA). In April 1940 he was made a temporary professor at the RIA, an appointment which helped his financial situation by coming with an annual salary of £1,000, and here he gave another series of lectures on quantum mechanics, this time at a more advanced level, again attracting sell-out audiences. Then, in May, he set off on a solitary cycling trip, catching the train to Galway before heading off into Connemara. It was there, just a couple of days into his holiday, that he learned from a newspaper that on 10 May the "phoney war" had ended with the German invasion of the Low Countries, the discovery sending him hurrying (rather pointlessly) back to Dublin.

The fall of France had little effect on life in the Irish capital, apart from the natural feelings of gloom engendered by Hitler's success, and there was nothing to stop the Schrödingers (just Erwin and Anny; some proprieties still had to be observed) accepting an invitation to a summer holiday with Paddy Browne and his sister's family at a house he owned on the Dingle Peninsula in County Kerry, with nothing except the wild Atlantic Ocean between them and America. Paddy's sister Margaret had three children. Her eldest, Máire (eighteen in 1940), later married the prominent Irish politician and writer Conor Cruise O'Brien. The middle child, Seámus, was sixteen at the time. And her youngest, a daughter called Barbara, was just twelve and made a big enough impact on Erwin that he had to be warned off by Paddy. Even this, though, did not affect their friendship.

Shortly after their return to Dublin, Schrödinger was at last able to take up his appointment at the Dublin Institute for Advanced Studies, with a salary of £1,200 a year. This was nearly half what the Taoiseach was paid, but once again Schrödinger was in a post which did not provide a widow's pension, although security was assured as long as he lived. He wrote to Max Born that "to be reinstated to absolute security (at least as regards yourself) at 53 by a foreign government, in my case fills you with—well, infinite gratitude."

The institute was housed in Merrion Square, a ten-minute walk from Trinity College Dublin (TCD), which had awarded Schrödinger an honorary doctorate on 3 July; a similar honour was conferred by the National University of Ireland on 11 July.[1] He was also made a member of TCD's Senior Common Room, a rare privilege for outsiders which Schrödinger treasured, often lunching at the college. The institute formally came into being on 5 October, under a council with Paddy Browne as chairman, and a governing board which included Schrödinger and McCrea. "The first I knew of it," McCrea told me, "was when I received a telephone call from Éamon de Valera in October 1940." Rather tickled by receiving such a call from the head of state of a neutral country while Britain was at war, he accepted de Valera's invitation to join the board, which had its first meeting on 21 November. As McCrea put it, Irish neutrality was very flexible, and there were no problems travelling between Northern Ireland and the South. One of the first acts of the board was to appoint Walter Heitler (1904–81), a refugee German physicist who had made a key contribution to the development of quantum chemistry, as an assistant professor. What McCrea described as "this splendid idea" must, he said, have come from Schrödinger himself. So the stage was set for the early success of de Valera's brainchild.

Early days at the DIAS

Schrödinger's "infinite gratitude" to the Irish encouraged him to work hard at making the institute a success, and because of his awareness that his security came at a cost to the Irish taxpayers he made a point of answering every letter he received from them, even the ones expounding crazy scientific "theories." The influence of the DIAS, which attracted eminent scientists from Britain to attend meetings (including visitors who had settled in the UK after fleeing the Nazis), extended to both the universities in Dublin, where members of the institute gave public lectures. The first scientific gathering at the institute (this one just for scholars based in Ireland) took place in the summer of 1941, at the time Heitler took up his post. As in Berlin in the old days, though, for the Schrödingers it was as important to play hard as to work hard. They gave tea parties and lunches at their house, especially for the younger members of the institute and students, whom Erwin always liked to encourage. By the end of 1940, there was a significant Austrian community in Dublin, and among these refugees was a young man, Alfred Schulhof, whose mother had been at school with Hilde. Schrödinger took Alfred under his wing, and paid for him to study electrical engineering. Erwin also "went native" sufficiently to attend the occasional cricket match.

The theatre was still one of Schrödinger's passions, and he attended plays accompanied by both his ladies. Moving in theatrical circles, they met the poet Patrick Kavanagh, the actress Sheila May, and her husband, David Greene, a Celtic scholar then working at the National Library in Dublin but later to join the Celtic studies school at the institute. The war came a little closer on 31 May 1941, when German bombs accidentally fell on Dublin, killing about thirty people; but three weeks later Hitler sealed his own fate by invading the Soviet Union. Schrödinger wrote in his diary: "It is really a great joy to see the two wretches [Hitler and Stalin] in battle against each other," but he clearly recognized that there could now be only one outcome to the conflict.

The following year, Schrödinger's own tolerant view of the world was shown during a storm in a teacup that surrounded a newspaper column written by the Irish humorist Brian O'Nolan, who used the pseudonym "Myles" (under which he wrote one of the great humorous novels, *The Third Policeman*). Schrödinger already knew Myles, through his circle of intellectual contacts, but that didn't persuade the columnist to pull any punches. Referring to a debate held at Trinity College in which Schrödinger had participated, Myles wrote in the *Irish Times*:

> I understand also that Professor Schroedinger has been proving lately that you cannot establish a first cause. The first fruit of the Institute, therefore, has been to show that there [is] no God. The propagation of heresy and unbelief has nothing to do with polite learning, and unless we are careful this Institute of ours will make us the laughing stock of the world.

This provoked a furious response from the council of the institute, which demanded an apology. But Schrödinger distanced himself from the argument, writing to the council: "I beg to decline emphatically the inclusion of any statement about my having been grieved by that article, or of any apology to me . . . or of anything that gives the wrong impression that I have asked for an apology." Although Schrödinger clearly was unconcerned, even amused, the council went ahead without him, and extracted a promise from the paper that Myles would never mention the institute again. But Myles and Erwin remained friends.

Shortly after this brouhaha, in the summer of 1942, a much more significant event in the history of the DIAS took place—its first international colloquium. To describe the meeting as "international" is a slight exaggeration, since most of the fifty or so people present came from Ireland, North and South. The most eminent of the Irish contributors was Ernest Walton (1903–95), born in County Waterford, who had carried out pioneering "atom-smashing" experiments in

Cambridge with John Cockcroft (1897–1967) and was now a Fellow of TCD. He became the only Irish scientist to receive a Nobel Prize for his work (shared with Cockcroft in 1951). But the main speakers were Dirac and Eddington, who each gave short lecture courses; McCrea, who was present, described the event as a "rare intellectual refreshment" in a time of war. Eddington was almost as shy an individual as Dirac, and McCrea recalled his surprise at seeing them both become unusually relaxed and sociable in the friendly environment provided by Schrödinger and his colleagues. McCrea always suspected that Schrödinger himself, who spoke perfect English when he wanted to, deliberately dropped non-standard idioms into his talks to help to get his meaning across and cut through formality, rather like Agatha Christie's Hercule Poirot (whom McCrea somewhat resembled). He thought it was a great shame that "orthodox" English was always used in the published versions of Schrödinger's lectures.

Soon after the colloquium ended, the Schrödinger household was increased as a result of a holiday Erwin and Anny took in Killarney. There they met a teenage girl, Lena Lean, whom they invited to join them in Dublin to help looking after Ruth; for once, this seems to have been a straightforward arrangement which can be taken at face value.

Schrödinger's own scientific work in the first couple of years of the institute's existence had concentrated mainly on developing the implications of Maxwell's electromagnetic theory, and this fed into his growing interest (almost an obsession) from 1943 onwards in the search for a single theory to unite gravity and electromagnetism. Although this quest was ultimately fruitless, he became an expert on the general theory of relativity. But something that started out almost as a relaxation from his main line of research would prove much more important and influential. The institute was required by its statutes to give public lectures every year, alternately at TCD and University College, and in February 1943 Schrödinger chose to give a series of three lectures himself, at TCD, on the way changes at a molecular level (in the genes) cause mutations expressed in the body plans of living organisms. These lectures, to which he gave the deliberately provocative title "What Is Life?," attracted an audience that included de Valera, senior members of the Irish Catholic Church, politicians, diplomats, and the intellectual élite of Dublin, as well as many ordinary folk. They were given on Fridays, starting on 5 February, but were so popular that Schrödinger had to repeat them on the following Mondays for the benefit of those unable to get in to the hall, which could hold four hundred people, on the Fridays.

The lectures, and the book which resulted from them, proved so important and influential that I shall tell the full story in the following chapter. But the key insight which he offered to a wide public is that chromosomes carry messages written in code, not unlike Morse code or, indeed, the alphabet in which the words you are reading are written. Not long after Schrödinger gave those lectures, his own life developed some more of its usual complications: before long, Ruth would have two half-sisters.

"Family" life in Dublin

During 1943, the marriage of David Greene and Sheila May ran into difficulties, and Sheila turned increasingly to her friend Erwin Schrödinger for consolation. Sheila was a feisty woman, actively involved in Irish politics as a member of the Labour Party, and engaged in a long-running battle with the authorities about the state of the slums in Dublin, where tuberculosis, rickets, and other diseases of malnutrition were rife. She bitterly pointed out that the only change that had been made to the slums since Ireland became independent was that the names of the narrow streets had been changed from English to Irish. Although Schrödinger was sympathetic to such views, theirs was an attraction of opposites, between Sheila the activist and Erwin the thinker. Their sexual affair began in the spring of 1944, when Erwin wrote in his diary: "What is Life? I asked in 1943. In 1944, Sheila May told me. Glory be to God!" As ever with Schrödinger, it wasn't just a physical thing, but love. A stream of poetry flowed from his pen, and in July he rented a flat in the centre of Dublin where they could meet. But the liaison was largely kept secret.

Although Sheila and David had by now been married for more than five years, they had no children because David did not want any. But soon, Sheila was pregnant, to Erwin's initial delight. He wrote: "I am the happiest man in Dublin, probably in Ireland, probably in Europe." But, as Walter Moore has put it in a memorable turn of phrase, "the mystic union of sexual love did not endure for long—with Erwin it was never able to survive tidings of pregnancy." By October, he was writing to Sheila to tell her to confess all to David and say that the affair was over. And David became the only person to emerge from the mess with any real credit: he not only accepted the child, a girl born on 9 June 1945 and christened Blathnaid Nicolette, but when he later separated from Sheila he had custody of her and brought her up as his own daughter.

In the spring of 1945, while Sheila was pregnant, Erwin met a young woman who worked with Hilde as a Red Cross volunteer, sending parcels to Austria via neutral Sweden. She is known to us as Kate Nolan—not her real name, because her family have always wanted to preserve their privacy. Kate was in many ways the opposite of Sheila. Although twenty-six when she met Schrödinger, she came from a strict Catholic background, had no intellectual pretensions, and was sexually inexperienced. It took Erwin some time to break down her resistance; he did so in the summer of 1945, and when the inevitable happened Kate confessed to Lena Lean, the Schrödingers' resident childminder, that she was not quite sure how she had become pregnant. Of all Schrödinger's "conquests," this is the hardest to justify on the grounds of "true love."

The baby, another girl, was born on 3 June 1946, and christened Linda Mary Therese; she was given the surname Russell in honour of Schrödinger's English antecedents. Kate's traditionally Catholic family were happy for her to be distanced from the child: Linda was unofficially adopted by the Schrödingers, and brought up in their household, with Lena's help. There was talk of making the adoption formal, and with the return of Hilde and Ruth to Arthur March's home in Innsbruck the household took on an almost regular appearance. But in 1948 Kate found Lena out walking with the baby in its pram, and took the infant away. She quickly put as much distance as she could between herself and Erwin, moving to southern Africa, and he never saw Linda again, although he always contributed to her maintenance, and gave £1,000 (the best part of a year's salary) to be invested for her for when she grew up. (There is another chapter to this story, which I tell in the Postscript to this book.)

While all this was going on, and Schrödinger was engaged in his unsuccessful pursuit of a unified field theory, Nazi Germany had been defeated on 7 May 1945, and the dropping of two nuclear bombs had brought an end to the war with Japan on 15 August. Schrödinger shared the distress of many physicists at the destruction wrought on Hiroshima and Nagasaki by the application of their craft—indeed, this reaction would be a major factor in the impact of his book *What Is Life?* Around the time the war ended, Heitler took over for a spell as Director of the Dublin Institute, with Schrödinger returning to the role once again in 1949, at which point Heitler left to follow in his footsteps as a professor at the University of Zürich. Thus Schrödinger had freedom to travel abroad as science across Europe and America returned to a peacetime footing.

The post-war years

The first benefit to Schrödinger's scientific life of the easing of travel came when Pauli visited Dublin from the Princeton Institute for Advanced Study in March 1946, bringing news of the latest developments in particle theory and nuclear physics (at least, those developments that were not still classified). More visitors followed, and then, in July, Erwin and Anny were able to go to England, visiting Cambridge to renew contact with Dirac, and (in Erwin's case) London to renew contact with Hansi. After five days with Hansi, Erwin linked up with Anny and they travelled on to Switzerland—first to Zürich, where Erwin gave a physics lecture, and then to Ascona to participate in a philosophical meeting devoted to "The Spirit of Nature," where Erwin spoke about "The Spirit of Science" to an audience that included Carl Jung. It seems a long way to go to preach the message that "the spirit, strictly speaking, can never be the object of scientific enquiry." The reaction of Jung, who spent his career trying to study the spirit scientifically, has not been recorded.

In 1946, Cambridge University Press published a short book by Schrödinger on statistical thermodynamics, based on a lecture course he had given at the Dublin Institute; but his main research interest in the next few years continued to be the search for a unified field theory. Some good did come of this—in 1950 CUP published another of his monographs, *Space-Time Structure*, which became a standard text for generations of students being introduced to the general theory of relativity.

Back in Ireland, during the campaigning for the Irish general election of 1947 (which de Valera's party lost), the institute came under criticism from some quarters as an expensive luxury that the country could not afford in what were then hard times; but its funding was secure, as was Schrödinger's future there, and on 17 February 1948 he and Anny became Irish citizens. The same month, he gave another series of public lectures, this time on "Nature and the Greeks," repeating them in London three months later, at University College. While he was in London, Schrödinger was introduced by Hansi to the potter Lucie Ray, who over the next few years would largely replace her in Erwin's affections. In the shorter term, his return to Ireland was to be followed by a series of personal problems that disturbed the calm of the Schrödingers' Dublin home.

Anny had always treated Ruth like her own daughter, and had become depressed in the months following her return to Innsbruck. Erwin's behaviour, and his infatuation with his new daughter, Linda, cannot have helped, and in June Anny made what was either a cry for help or a genuine suicide attempt by slashing her wrists. She spent several weeks in St. Patrick's Hospital, where she was given electric shock therapy, then a routine treatment for psychological disturbance, which seemed to help. But she suffered from recurrent bouts of depression through the rest of the Schrödingers' time in Dublin, and admitted herself to hospital for treatment several times. The situation was not helped by her need to take steroids to control her asthma, which made her put on weight and feel unattractive.

Erwin's problem was more easily treated. He had been developing cataracts in both eyes, and on 29 June (while Anny was in hospital) had an operation on his right eye. The cataract in his left eye was removed the following year. Both operations were completely successful. But it was not long after the first of these operations and Anny's discharge from hospital that Linda was snatched by her mother. None of this seemed to disturb Erwin's equanimity, whatever its effect on Anny. In August he went on holiday to North Wales with Hansi, staying for a while in Portmeirion, where they bumped into Bertrand Russell; in September he attended the eighth Solvay Congress in Brussels, devoted to the subject of "Elementary Particles"; and back in Dublin he still spent most Saturdays hiking with some of his colleagues in the Wicklow Mountains.

In May 1949, Schrödinger received an overdue honour—he was elected as a foreign member of the Royal Society. Why had it been so long coming? In a word: "politics." Between 1938 and 1948 no German or Austrian citizens were elected to the Royal. In addition, before 1948 only foreign nationals who were not resident in one of the British Dominions were eligible for election. Technically, Ireland was still a British Dominion until 1948, when the new Irish government formally severed this last link with the past. Although the new Fellow of the Royal Society did not publish any scientific papers in 1949 (his first "fallow" year since 1923), he did issue a slim volume of poetry, which would surely not have appeared were it not for his fame as a physicist. Schrödinger's poetry, in fact, reads almost like a pastiche of the kind of poetry you would expect a physicist to write—it is technically correct, in terms of metre, rhyme, and so on, but lacks the emotional impact of the work of a true poet. Much more interesting was his recording, the same year, of a talk at the BBC in London for a series on "Frontiers of Science"; the recording still exists, and confirms the accuracy of McCrea's comment that Schrödinger could speak perfect English when he chose to. Two more talks, recorded in 1950, were broadcast only on the European Service of the BBC.

In the first quarter of 1951, at the recommendation of Arthur March, Schrödinger spent a term at the University of Innsbruck. With Austria still occupied by the victorious wartime Allies, the situation was (as I shall explain in Chapter 13) politically complicated, but Innsbruck was in the French zone, where things were relatively calm. The trip did more than just rekindle old memories; it provided Erwin with an opportunity to meet up with his daughter Ruth, who was now sixteen (although it was another year or so before she learned that he was her father). He gave a short series of lectures in

Vienna on the general theory of relativity, and was sounded out about the possibility of a permanent post in Innsbruck. Although in the end the university was unable to find room for him, in the years that followed Schrödinger often returned to Austria, attending summer conferences at Alpbach and extending these into holidays.

During Schrödinger's remaining years in Dublin, after he largely abandoned the search for a unified field theory as a dead end, he did little scientific work that received recognition at the time, but gave many lectures, and wrote some papers, on the fundamental problem of the interpretation of quantum mechanics. These have largely been ignored, regarded as minor works of his scientific dotage, looking backward rather than forward and tidying up loose ends. But they actually contain profound insights, very relevant today, both into his thinking and into the nature of the world. These deserve proper attention. I shall come on to this shortly, after briefly summarizing the key events in Schrödinger's life up until his return to Vienna.

In August 1952 Erwin reached his sixty-fifth birthday. After a holiday in the Tyrol the following month he was feeling full of vigour, and looking forward both to the new academic year in Dublin and to a conference on the interpretation of quantum mechanics to be held in London that December. All the old protagonists would be there for a rerun of the debate between Bohr and his fellow "Copenhagers" and the rest. Schrödinger wrote a paper to contribute at the conference, but he never got to present it in person. At the end of October he was struck down with appendicitis. The appendix burst, and he needed emergency surgery; his life was probably only saved by the availability of the antibiotics that were just coming into general use. But he never fully recovered the physical well-being he had felt just a month before, and the attacks of bronchitis he suffered every winter became more severe—not that they stopped him from smoking his pipe as much as ever. By 1954, on his regular summer visit to the Tyrol he had so much difficulty keeping up with his companions on long walks that he was persuaded to see a doctor, who diagnosed severe emphysema and high blood pressure. It later became clear that he was also suffering from arteriosclerosis. From then on, alcohol was forbidden, smoking was restricted, his bedtime was 9 p.m., and hiking in the mountains became a thing of the past.

In December 1952 Schrödinger was asked to spend a semester at Harvard, and initially accepted the offer, making arrangements with the institute for an absence from October to December 1954. But when the Harvard authorities changed the dates to run from 25 September until the end of January 1955, and said his duties would include marking examination papers, he changed his mind. There was only one place that Schrödinger was really willing to leave Dublin for—Austria. The opportunity arose, as I describe in Chapter 13, in 1956. But first, it is time to look at Schrödinger's lasting scientific legacy from his time in Dublin, starting with the interpretation of quantum mechanics.

Many worlds

Schrödinger's lecture notes for his seminars, and other unpublished material from his later years in Dublin, were eventually gathered together and edited by Michel Bitbol, of the Centre National de la Techerche Scientifique in Paris; they appeared in book form in 1995. Bitbol explored the underlying philosophy behind this work in another book, published a year later, *Schrödinger's Philosophy of Quantum Mechanics*. He makes a convincing case that this body of work is the culmination of Schrödinger's thinking about quantum mechanics, and presents a view very different from the Copenhagen Interpretation, but very close to a modern understanding.

The basis for Schrödinger's thinking is actually summed up in one of his published papers, which appeared in 1952 with the title "Are There Quantum Jumps?" He said that there is nothing in experimental physics which has to be interpreted in terms of discrete particles (remember the example of the trail of dots left by an electron in a cloud chamber). We may not be sure what a "particle" is, said Schrödinger, but "we have now gained [insight] into what it is not; it is not a durable little thing with individuality." All we have from experiments are the records of events, which we examine long after the events have happened—an observation that is, if anything, even more true today of experiments involving machines like the Large Hadron Collider than it was for the simpler experiments of Schrödinger's day. If we see an electron in position *A*, and later (even a split second later) see an electron in nearby (even very nearby) position *B*, we have no way of knowing whether it is, in fact, the same electron. And particles which have neither well-defined trajectories nor well-defined individuality simply are not particles. "It is better to regard a particle not as a permanent entity but as an instantaneous event," said Schrödinger.[2] "Sometimes these events form chains that give the illusion of permanent beings."

This also removes, at a stroke, the puzzle of action at a distance that so worried Einstein. Instead of thinking of two separate particles, with two separate but entangled wave functions that interact with one another in some spooky fashion, we should think of a single wave function which describes the whole system. In terms of "local reality," we can keep reality if we abandon locality.

I have already mentioned the next, key, step in the argument. There is nothing in the equations that requires the collapse of the wave function. In the "cat in the box" thought experiment, the Copenhagen Interpretation tells us that there is a superposition of wave functions, or states, until the box is opened. Then, the system collapses and just one state becomes "real." Why? asked Schrödinger. There is no reason for the superposition to be disturbed just by our looking at it—and remember that in 1935 he had said that the superposition of states is "the characteristic [trait] of quantum mechanics" (see note 3 to Chapter 9). Some fifteen years later, he said: "It is patently absurd to let the wave function be controlled in two entirely different ways, at times by the wave equation, but occasionally by direct interference of the observer, not controlled by the wave equation."

Now, there is in any case something odd about the superposition of wave functions in quantum mechanics. In the realm of classical waves, two superimposed waves add together to produce a composite wave; they do not retain their separate identities. So wave mechanics is a radical departure from classical wave theory, and not, as was originally hoped and is still often supposed, less radical than matrix mechanics. By the early 1950s, Schrödinger is telling us that in the cat experiment both states are real, and they each stay real after the box is opened. The mind-boggling implication is that all quantum states are real. This is the foundation of what later became known as the "Many Worlds" interpretation of quantum mechanics, although nobody before Bitbol seems to have noticed that Schrödinger thought of it first. The key passage comes from a talk Schrödinger gave in Dublin in 1952:

> Nearly every result [the quantum theorist] pronounces is about the probability of this or that or that . . . happening—with usually a great many alternatives. The idea that they may not be alternatives but all really happen simultaneously seems lunatic to him, just impossible. He thinks that if the laws of nature took this form for, let me say, a quarter of an hour, we should find our surroundings rapidly turning into a quagmire, or sort of a featureless jelly or plasma, all contours becoming blurred, we ourselves probably becoming jelly fish. It is strange that he should believe this. For I understand he grants that unobserved nature does behave this way—namely according to the wave equation. The aforesaid alternatives come into play only when we make an observation—which need, of course, not be a scientific observation. Still it would seem that, according to the quantum theorist, nature is prevented from rapid jellification only by our perceiving or observing it . . . it is a strange decision.

Rather than undergoing "jellification," though, with no collapse of the wave function the cat in the box becomes two cats in two boxes, in two separate branches of the world (or two separate worlds); one cat dies, the other lives. And here is a multitude of branches of reality—many worlds—corresponding to every possible quantum state of the Universe. This removes the puzzle of who has to observe the Universe itself to make it collapse into a definite state.

The most strenuous objection to this idea has come from people who cannot stomach the idea of the Universe constantly splitting into new versions of itself every time it is confronted by a quantum "choice." As I shall describe in Chapter 14, this objection has now been overcome, although Schrödinger did not live to see it happen. But he did live to see widespread recognition of another piece of his later work—his ideas about the nature of life.

Chapter Twelve

What Is Life?

Schrödinger had a lifelong interest in the process of heredity, having learned about biology from his botanist father and read widely about evolutionary ideas during his time as an undergraduate, when the recently rediscovered work of Gregor Mendel (1822–84) on inheritance was being widely discussed. His interests in philosophy and Eastern religion, addressing the nature of the mind and soul, and questions such as the possible existence of a group unconscious, formed part of the same tapestry of thought. To Schrödinger, the continuity of the genetic line is a kind of immortality, and he always regretted never having a son. So when the time came around for another series of public lectures, in 1943, Schrödinger decided to treat his audience to his thoughts about the nature of life and inheritance, jumping off from a paper co-written by Max Delbrück (1906–81), whom Schrödinger knew from his time in Berlin, when Delbrück was working at the Kaiser Wilhelm Institute for Chemistry.

Although the two may have discussed Delbrück's work in the early 1930s, the key paper was not published until 1935, after Schrödinger had left Berlin, and then in a relatively obscure journal. Schrödinger probably saw it only a short time before he decided to make this work the basis of his public lectures at TCD in February 1943.

Today "everybody knows" about DNA, and the term "genetic code" has become part of the common language. It is difficult to stand back and imagine the impact of the ideas of Delbrück, as interpreted by Schrödinger, in the 1940s, and ridiculous to try to write about these events without acknowledging how much more we now know. So, to put the impact of Schrödinger's book *What Is Life?* in perspective it seems best to get a clear idea first of just what it is that we do now know about DNA, the genetic code, and inheritance.

Life itself

DNA is a long molecule found in the cells of every living thing. The most important feature of DNA is that strung out along the length of the molecule there is a series of chemical subunits called bases, which are denoted by the letters A, C, T, and G. Strings of these four bases can convey information in what is usually called a code, but which I prefer to think of as a language, in the same way that the twenty-six letters of the alphabet are used in long strings to convey information in this book. But DNA molecules do not usually exist in isolation. They come in pairs, with one long molecule twined around its partner in the famous double helix arrangement. The two molecules in each helix are not identical, but are like mirror images of one another; everywhere one molecule has an A, its partner has a T; everywhere one molecule has a C, its partner has a G; everywhere one molecule has a G, its partner has a C; everywhere one molecule has a T, its partner has an A. So under the right circumstances (which occur when a living cell divides) the two halves of the helix can unwind, and each single strand can build itself a new partner from the chemical material surrounding it inside the cell, by making appropriate links between bases. The result is two identical double helices, and one copy goes into each of the two cells formed by the division.

When sex cells (sperm or egg) are being manufactured by the body, a slightly more complicated process occurs in which pieces of DNA get cut out of one helix and spliced into another, so that offspring inherit a slightly different arrangement of genetic material from their parents.

All of this matters because the code or language carried by the DNA contains the instructions for the construction of an organism from a single cell, and for the operation of the organism. The code/language is translated and the instructions are put into action inside a living cell with the aid of a molecule very similar to DNA, called RNA. In the process, a section of DNA double helix untwists, and the relevant section of code is copied into a single-stranded RNA molecule. This RNA "message" is then used, by the machinery of the cell, to build up molecules called amino acids, which are linked up to form proteins. Some proteins provide the structure of your body, things like muscle and hair, while others, known as enzymes, act as catalysts, encouraging (or in some cases inhibiting) chemical reactions going on inside your cells.

Proteins are so important in the body that in the early 1940s, when Schrödinger was writing his book, they were widely thought to be the molecules of life, and DNA was thought to be merely some kind of scaffolding on which protein chemistry could take place, without contributing directly to the process. But it was known that the genetic information is packed into entities called chromosomes—each individual human has 23 pairs of chromosomes, with one member of each pair inherited from each parent (it is actually chromosomes that are broken apart and joined together in new arrangements when sex cells are made). Genes are sections of chromosomes, and it is a change in a gene (sometimes called a mutation) that produces changes in individual members of a species on which evolution can act. But how big does a change in the molecule of life (whatever it may be) have to be in order to produce a significant change in the individual? In the 1935 paper that so intrigued Schrödinger, Delbrück and his colleagues, using data from experiments in which mutations were caused in fruit flies (drosophila) by X-rays, suggested that a mutation can be caused by a single change at one place in a molecule—in modern terminology, a change as simple as changing an A to a G in a DNA helix. The scientific paper that conveyed this dramatic information became known, from the colour of the cover on the reprints that circulated (increasingly after Schrödinger drew attention to it), as "the green pamphlet."

But just as Schrödinger's *What Is Life?* drew on the green pamphlet, so Delbrück and his colleagues drew on Schrödinger's earlier work, since the kind of biology they were concerned with is part of chemistry, and by the 1930s chemistry had become part of physics—specifically, part of quantum physics. And the version of quantum physics that chemists used was Schrödinger's wave mechanics.

Quantum chemistry

Chemistry is concerned with the way atoms join together to make molecules. This involves a sharing of electrons (which have negative charge) between the positively charged nuclei of different atoms. In the simplest example, two atoms of hydrogen, which each consist of a single proton associated with a single electron, join together to make a single molecule of hydrogen, in which in some way the two electrons "surround" both of the protons (the atomic nuclei). Molecules form in this way because the molecular state represents a lower-energy state than the two atoms on their own. But how could two electrons surround two protons? It's like saying that two small children "surround" their parents.

Clearly, it is much easier to visualize this sharing of electrons, in which a couple of electrons can in some sense surround a pair of atomic nuclei, using the idea of waves rather than that of particles. This does beg the question of how the electric charge associated with the electrons gets "smeared out," and leads straight to Born's idea that the wave represents a probability of finding the electron in a particular place, but that the electron actually exists as a particle—the interpretation that Schrödinger hated. For the purposes of chemistry, though, such puzzles can be left to the philosophers and interpreters of quantum mechanics. The chemists' concern, once Schrödinger had discovered wave mechanics, was to find ways to use the equations to calculate the change in energy that occurs when hydrogen atoms combine to make molecules, and then to extend this kind of calculation to more complicated systems, so that they could predict which arrangements of atoms would form stable molecules, and how strong the bonds between them would be.

The theory of the electron bond was developed, using Schrödinger's wave equation, from the independent work in 1927 of the American Edward Condon (1902–74) and the team of Walter Heitler and Fritz London. That summer of 1927 Heitler and London were in Zürich, and benefited from discussions with Schrödinger both in the formal setting of the university and during long walks in the woods. Linus Pauling (1901–94), who was to become the leading quantum chemist, was also in Zürich that summer, but had little contact with Schrödinger.[1] When Heitler and London calculated the difference in energy between two hydrogen atoms and one hydrogen molecule, they came up with a value very close to the amount of energy which chemists already knew, from experiment, was required to break such a molecule apart. This was a dramatic discovery, since it showed that the arrangements of atoms in molecules are not arbitrary, but are indeed the arrangements with least energy, and therefore the most stable arrangements.

It was Pauling who developed a complete, coherent, and above all quantitative description of the chemical basis of biology over the next few years, in particular using wave mechanics to explain the chemical behaviour of carbon, the single most important atom in the chemistry of life—so important for life that the terms "carbon chemistry" and "organic chemistry" are synonymous. His book *The Nature of the Chemical Bond*, published in 1939, became the most influential chemistry textbook of the twentieth century. In 1954 Pauling received the Nobel Prize for this work, with the citation specifying "his research into the nature of the chemical bond and its application to the elucidation of the structure of complex substances." But its influence had begun to spread well before the book appeared; indeed, his research into the nature of the chemical bond had essentially been completed by 1935—which is why the authors of the green pamphlet were able to relate the energy carried by individual X-ray photons to the energy required to break different kinds of chemical bond.

The green pamphlet

In fact, the section of the green pamphlet that discussed the X-ray experiments was written by Delbrück's colleague Nikolai Timofeev-Ressovsky (1900–81). The "third man" was a more junior colleague, Karl Zimmer (1911–88), who used the data from the experiments to calculate how much energy is needed to cause a mutation, and, by comparing this with the bond energies of carbon compounds calculated using quantum chemistry, concluded that a single "hit" by an X-ray photon could produce a mutation. It later turned out that some of the assumptions in his calculation were wrong; but fortunately this did not affect the main conclusion. This was that a chemical change involving at most a few hundred atoms and perhaps equivalent to the breaking of a single molecular bond (followed by the formation of a new and different bond) could produce a genetic mutation. This was one of the first pieces of evidence that genes are, indeed, molecules, and not more complicated structures like tiny versions of the cell; it led first Delbrück, in his section of the green pamphlet, and then Schrödinger to the idea of the genetic code.

Max Delbrück had been born in Berlin in 1906 and worked his way through the German educational system, studying mathematics, physics, and astrophysics. He emerged with a PhD in 1930 and thereafter spent short spells in

England at the University of Bristol's H. H. Wills Physics Laboratory, with Niels Bohr in Copenhagen, and with Wolfgang Pauli in Zürich, before settling at the Kaiser Wilhelm Institute for Chemistry in Berlin. He was one of the first physicists to make the move through chemistry into biology in the wake of the development of quantum mechanics; this trend did not gain pace until after the Second World War. In 1935 Delbrück himself was working as an assistant to the physicist Lise Meitner; his biological interests were at that time still secondary to his main research, although three decades later he would receive a share of the Nobel Prize for his work on the genetics of viruses. The impact of the new physics on biology is, though, neatly encapsulated in the title of Delbrück's section of the pamphlet—"A Model of Mutation Based on Atomic Physics."[2]

The direct inspiration for this work was a lecture Delbrück attended, entitled "Light and Life," given by Bohr in Copenhagen in 1932 and published in *Nature* the following year. Bohr said:

> The existence of life must be considered as an elementary fact that cannot be explained, but must be taken as a starting point in biology, in a similar way as the quantum of action, which appears as an irrational element from the point of view of classical mechanical physics, taken together with the existence of elementary particles, forms the foundation of atomic physics. The asserted impossibility of a physical or chemical explanation of the function peculiar to life would be . . . analogous to the insufficiency of the mechanical analysis for the understanding of the stability of atoms.

But Bohr insisted that there was no need to invoke a mysterious "life force" to explain the difference between living and non-living things. "If we were able to push the analysis of the mechanism of living organisms as far as that of atomic phenomena," he argued, "we should scarcely expect to find any features differing from the properties of inorganic matter." It was the idea of pushing the analysis of the mechanism of living organisms to this limit that led Delbrück into genetics. One of the reasons why he was attracted to the job with Meitner in Berlin was that the Kaiser Wilhelm Institutes were set up in close proximity to one another, to encourage the spread of ideas across disciplines; Berlin in the early 1930s was the ideal place for a physicist who was interested in biology.

In the green pamphlet, Delbrück suggested that genetic mutations result from the transitions of molecules from one quantum state to another. Genes must be very stable molecules, he pointed out, in order to transmit characteristics from one generation to the next. Rare mutations, the kind on which natural selection acts, can occur as a result of the molecules absorbing energy from their surroundings, perhaps simply through the action of heat energy jostling the molecules about; more frequent mutations can be caused by adding energy in the form of X-rays or (as Delbrück speculated but was later proved) ultraviolet radiation. It is the rearrangement of the genetic material (whatever it may be) in quantum processes that causes mutations. In other words, mutation is a quantum process that involves molecules being pushed from one stable configuration to another stable configuration over an energy barrier. But what are the genes? It was too soon to say, and in Delbrück's words:

> We leave open the question whether a single gene is a polymeric entity that arises by the repetition of identical atomic structures or whether such periodicity is absent; and whether individual genes are separate atomic assemblies or largely autonomous parts of a large structure, i.e. whether a chromosome contains a row of separate genes like a string of pearls, or a physico-chemical continuum.

A polymer is simply a long molecule containing very many atoms strung out like a—well, like a string. Delbrück's introduction of the idea of the gene as a polymeric molecule is a key step in the development of an understanding of the mechanisms of life. You shouldn't be confused, though, by his reference to "the repetition of identical atomic structures"; if they were all literally identical then they would not convey any information, just as the string of letters AAAAA . . . conveys no information, and he clearly means the repetition of a few identical units, but in different orders, like different words written using the same letters of the alphabet.

As a result of the work described in the green pamphlet, Delbrück was awarded a Rockefeller fellowship which took him to California in 1937, initially to work with the geneticist Thomas Hunt Morgan (1866–1945) at Pasadena. He remained based in the United States for the rest of his life, and became an American citizen in 1945. While in California, he also worked with Linus Pauling, and together they wrote a paper, published in the journal *Science* in 1940, in which they pointed out that two molecules with complementary structure (using a term clearly borrowed from quantum mechanics, but meaning the same as my use of the term "mirror image molecules") lying alongside each other would form a particularly stable configuration. They said that studies of this form of complementarity should receive a high priority in the investigation of the workings of the cell.

Pauling and Delbrück seem to have been unaware, at the time they wrote their paper, of a suggestion made in 1937 by the British geneticist J. B. S. Haldane (1892–1964), then based at University College in London. He said: "We could conceive of a [copying] process [of the gene] analogous to the copying of a gramophone record by the intermediation of a negative, perhaps related to the original as an antibody is to an antigen."[3] This is exactly the kind of process that, as I explained earlier, operates when DNA is copied in the cell, or when its message is translated using RNA. It matches up beautifully with the suggestion made by Pauling and Delbrück. But Schrödinger seems to have been unaware of either of these suggestions when he wrote *What Is Life?*

Schrödinger's variation on the theme

The heart of Schrödinger's book is a reworking of the ideas from the green pamphlet, presenting the quantum-mechanical evidence followed by a chapter titled "Delbrück's Model Discussed and Tested." He says that it has "often been asked" how the tiny speck of material that is a fertilized egg "could contain an elaborate code-script involving all future development of the organism." His answer is that "the number of atoms in such a structure need not be very large to produce an almost unlimited number of possible arrangements," and he gives the example of a kind of super Morse code, with three signs instead of the usual dot and dash, which, used in groups of not more than ten, "could form 88,572 different 'letters';

with five signs and groups up to 25 the number is 372,529,029,846,191,405."

There is also a biological example that Schrödinger might have used, if he had been a biologist himself, and familiar with the latest work. By the end of the 1930s, it had been found that all proteins in living things are made up from different arrangements of just 20 different amino-acid building blocks. There are potentially about 24×10^{17} (24 followed by 17 zeroes) ways of arranging the "letters" in a 20-character "alphabet," meaning that there are potentially 24×10^{17} different possible proteins. Only a tiny fraction of these potential proteins are actually expressed in living things.

The gene, said Schrödinger, is like "an aperiodic crystal." He points out that in an ordinary (periodic) crystal of a substance such as common salt there is an endless repetition of the same basic unit in a perfectly regular pattern that conveys very little information, and contrasts this with "say, a Raphael tapestry, which shows no dull repetition, but an elaborate, coherent, meaningful design," even though it is made up of a few simple, identical units (in this case, different-coloured threads). This idea of an aperiodic crystal is essentially what Delbrück meant by "a polymer that arises by the repetition of identical atomic structures," but no doubt Schrödinger wanted to use a different analogy. And he certainly put the message across forcefully:

> In calling the structure of the chromosome fibres a code-script we mean that the all-penetrating mind, once conceived by Laplace, to which every causal connection lay immediately open, could tell from their structure whether

> the egg would develop, under suitable conditions, into a black cock or a speckled hen, into a fly or a maize plant, a rhododendron, a beetle, a mouse or a woman. To which we may add, that the appearances of the egg cells are very often remarkably similar; . . .

But the term "code-script" is, of course, too narrow. The chromosome structures are at the same time instrumental in bringing about the development they foreshadow. They are law-code and executive power—or, to use another simile, they are architect's plan and builder's craft—in one.

Apart from promoting the idea of a genetic code (without, it has to be said, offering any explanation of how the code might be copied), Schrödinger's book was important because it introduced the idea that life feeds off "negative entropy." This line of argument drew on Schrödinger's long-term interest in thermodynamics, which tells us that the entropy of a closed system always increases—that ordered systems become disordered. Life is clearly working the other way; it seems to create order out of disorder, or in Schrödinger's words it "evades the decay to equilibrium." Schrödinger seems to suggest that living things take in "negative entropy" from their surroundings in the form of food, which is in an ordered state. "The essential thing in metabolism is that the organism succeeds in freeing itself from all the entropy it cannot help producing while alive."

I say he "seems to" suggest this because this section of his book is very confused, and has drawn a lot of criticism. The obvious question is, where did the negative entropy in the food come from? But in spite of the confusion, Schrödinger was pointing in the right direction. Life does feed off negative entropy, which comes from the Sun; overall, the entropy of the solar system is increasing, in line with the laws of thermodynamics. A small decrease in entropy associated with the presence of life on Earth is vastly offset by the huge increase in entropy associated with the way heat from the Sun flows out into the cold of space, warming the Earth on its way. The key point is that the Earth is not a closed system in the thermodynamic sense. In fact, Boltzmann himself said nearly the same thing as far back as 1886, when he referred to the need of biological organisms "for entropy which becomes available by the transition of energy from the hot sun to the cold earth."[4] Confused (or not) though Schrödinger's discussion may have been, the term "negative entropy" caught on—when his book eventually got published.

The path to publication was complicated because when he turned his lecture material into book form Schrödinger added an Epilogue, "On Determinism and Free Will," in which he drew on Eastern religion and philosophy to express his version of the idea that the individual, personal self is only a facet of the universal self. "It is daring," he said, "to give to this conclusion the simple wording that it requires. In Christian terminology to say: 'Hence I am God Almighty' sounds both blasphemous and lunatic." He was right—it did sound blasphemous, and was hardly a sensible thing to say in a book being published in Dublin in the 1940s.

Surprisingly, perhaps, the book had got past the stage of being typeset (this was in the days of metal type set out by hand, letter by letter) and was in page proof with the publisher Cahill & Company before anything hit the fan. At that point, somebody reading the proof (almost certainly Paddy Browne, who had been helping Schrödinger turn his lecture notes into the kind of English thought appropriate for a book) went ballistic and drew the attention of the managing director at Cahill, John O'Leary, to the Epilogue. O'Leary refused to publish the book unless the Epilogue was removed; Schrödinger refused to remove the Epilogue; the type was literally broken up; and eventually (in 1944) the book, complete with Epilogue, was published by Cambridge University Press—ironically, a much better home for the book, helping to ensure that it was widely read.

The double helix

What Is Life? has often been criticized (accurately) on the grounds that what is good in the book is not original, and that what is original in the book is not good. This misses the point that you do not have to be original to be influential, and influential the book certainly was—no passage more so than this: "From Delbrück's general picture of the hereditary substance it emerges that living matter, while not eluding the 'laws of physics' as established up to date, is likely to involve 'other laws of physics' hitherto unknown, which, however, once they have been revealed, will form just as integral a part of this science as the former." The prospect of new applications of physics was immensely appealing to a whole generation of physicists, wearied by war and in many cases deeply concerned about the part the old physics had played in what happened to Hiroshima and Nagasaki. Working for life had more appeal than working for death.

Nothing shows this upsurge of enthusiasm—and Schrödinger's influence—more clearly than a conference which took place under the auspices of the US National Academy of Sciences in Washington, DC, in the autumn of 1946, just about as soon as such a conference could have been organized after the end of the war. The title was "Borderline Problems in Physics and Biology," and in the opening address to the meeting Max Delbrück noted that *What Is Life?* had been the stimulus that had brought them together. It is natural that physicists should have had their eyes opened to the biological possibilities of their work by Schrödinger's book; but from a modern perspective it looks at first sight rather more surprising that it had an equally profound impact on biologists. You have to remember that in the mid-1940s, only a few years after the publication of Pauling's great book, and with war intervening, very few biologists knew much (if anything) about the nature of the chemical bond or thermodynamics. This is not the place to go into all the details of how the message spread, but it is surely worth looking at the direct influence on the two scientists credited with discovering the structure of DNA, Francis Crick and James Watson.

Crick, who was born in a village near Northampton in 1916 (he died in 2004), was the senior member of the team in terms of age. He graduated from University College in London with a degree in physics in 1938, but his plans to carry out research for a physics PhD were interrupted by the war, when he worked for the Admiralty (up until 1947, in fact) on the design of acoustic and magnetic mines. His interest in the application of physics to biology was fired first by Schrödinger's book; "it was only later," he wrote in his memoir *What Mad Pursuit*, "that I came to see its limitations—like many physicists, he knew nothing of chemistry—but he certainly made it seem as if great things were just around the corner." The flames of this interest were fanned by an article written in the journal *Chemical and Engineering News* by Linus

Pauling in 1946. In 1947 Crick began working at the Strangeways Laboratory in Cambridge on a study of the way magnetic particles move in cells, and in 1949 he transferred to the Cavendish Laboratory to begin, at the late age of thirty-three, research for a PhD, involving X-ray studies of proteins.

It was at the Cavendish that Crick met Watson. Watson had been born in Chicago in 1928 and graduated from the university there, having studied zoology, at the age of nineteen. Crick moved from physics into biophysics; Watson moved from biology into biophysics. He had also read Schrödinger's book, in 1946 while still an undergraduate, and it was instrumental in determining his future career path. He said in 1984, in a talk given at Indiana University: "From the moment I read Schrödinger's *What Is Life?* I became polarised towards finding out the secret of the gene." With typical chutzpah, he also said: "It was clear in those days that physicists were brighter than biologists." Although he started working for a PhD on drosophila at Indiana University, in Bloomington, he soon switched to X-ray studies of a type of virus known as a bacteriophage. Armed with a fresh PhD and still only twenty-two, in 1950 Watson travelled to Copenhagen, where he carried out more work on bacteriophage, and then, in 1951, to Cambridge, where he met up with Crick at the Cavendish—through the pure chance of their happening to share a room at the laboratory.

At that time, the idea that biological molecules might have helical structure was in the air. In 1951, Pauling's team published an astonishing series of seven scientific papers describing the structure of many different proteins (in things like hair, feathers, muscle, silk, and horn) in terms of a structure known as the alpha-helix. By that time, it was also clear that the important molecules in chromosomes were in the form of DNA, and Watson, in particular, was inspired by Pauling's work to look for a helical structure in DNA, recruiting Crick to the cause and deflecting him from his protein work. The way to find such a structure was using X-ray crystallography, a technique with which neither Crick nor Watson had any expertise. But at King's College in London Rosalind Franklin (1920–58) was carrying out exactly this kind of study of DNA. As is now very well known, Crick and Watson obtained some of Franklin's key data by not entirely ethical means, and used this material as the basis for their famous model of DNA as a double helix.[5] The discovery was published in the issue of *Nature* dated 25 April 1953 (a year before Crick obtained his PhD); by the time it was recognized by the award of a Nobel Prize, in 1962, Franklin was dead (from cancer, possibly related to her work with X-rays) and could not share the honour.

Crick and Watson were well aware of the importance of their discovery for an understanding of the way DNA molecules (and therefore genes) replicate. Near the end of their paper, they wrote: "It has not escaped our notice that the specific pairing we have postulated immediately suggests a possible copying mechanism for the genetic material." But the discovery of the structure of DNA, and even the hint of how it replicates, was not the end of the story. The question still remained of how the information was coded in the DNA molecules that made up the genes in chromosomes, and how the information in the genes was transferred and put to use by the mechanisms of the cell. Crick would go on to play a key part in cracking the DNA code.

Although he was delighted by the impact his book had in the 1950s, Schrödinger didn't quite live to see the completion of this vindication of the ideas presented in *What Is Life?*: he died in January 1961, a few months before the news broke that Crick had broken the code—just five years after his long delayed return to Vienna.

Chapter Thirteen

Back to Vienna

The complications that delayed Schrödinger's permanent return to Austria resulted from the antipathy between the Western Allies in the war against Hitler's Germany and the Soviet Union. United only by their opposition to the Nazis, East and West were soon embroiled in a Cold War, neither willing to give up territory to the other. Like Germany, Austria was divided into separate zones of occupation after the war, with the Americans, British, French, and Russians each responsible for part of the country; and like Berlin, Vienna was similarly divided, as vividly portrayed in the film *The Third Man*. It took until 1990 for Germany to be reunified, and from that perspective Austria was lucky. But it didn't feel like a lucky country in the years immediately after 1945.

In some ways, the immediate aftermath of the war was like a replay of the aftermath of the First World War. Severe winters, especially in 1946–47, brought food shortages and street riots. There was an unsuccessful Communist coup attempt in May 1947, and in February 1948 the Communists seized power in neighbouring Czechoslovakia. Anticipating that Austria would be the next domino to fall, Josef Stalin, the Soviet leader, stalled the negotiations for an Austrian peace treaty. But there was a vital difference between the situation in 1948 and that of 1918. Instead of punishing Austria more severely, after the attempted coup of May 1947 aid (in the form of the Marshall Plan) poured in, as America sought to bolster this bulwark against the advance of communism. One side-effect of this was that in a spirit of reconciliation, outside the Soviet zone many former Nazis were allowed to remain, or return to, their posts in the civil service and academia; this was not possible for the former Jewish population, 90 per cent of whom had been murdered or exiled.

It was against this background that Schrödinger became a regular visitor to Austria in the early 1950s. He even managed to lecture in Vienna, although not without, on at least one occasion, some harassment by border guards while getting out of the Soviet zone (getting into the Soviet zone was easy!). With Anny's bouts of depression, and Erwin's declining health, it had begun to look as if they might end their days in Dublin after all. But in March 1953 Stalin died, and his successor, Nikita Khrushchev, soon decided that he had nothing to lose, and might gain politically, by clearing the air over Austria. The basis for the new treaty lay in an agreement known as the Moscow Declaration, signed during the Moscow Conference of October 1943 by the foreign ministers of the UK, the USA, and the USSR.

The section of the declaration regarding Austria stated that the annexation of Austria by Germany was null and void, and called for the establishment of a free Austria after the victory over Nazi Germany:

> The Governments of the United Kingdom, the Soviet Union and the United States of America are agreed that Austria, the first free country to fall a victim to Hitlerite aggression, shall be liberated from German domination.
>
> They regard the annexation imposed upon Austria by Germany on March 15, 1938, as null and void. They consider themselves as in no way bound by any changes effected in Austria since that date. They declare that they wish to see re-established a free and independent Austria, and thereby to open the way for the Austrian people themselves, as well as those neighbouring states which will be faced with similar problems, to find that political and economic security which is the only basis for lasting peace.

The painstaking negotiations over a peace treaty still went on for months after Khrushchev's decision to proceed, but it was duly signed on 15 May 1955; it officially came into force on 27 July, and by early November the occupying forces had left Austrian soil. In between, Schrödinger was appointed to a professorial post specially created for him at the University of Vienna—receiving the news while on one of his regular summer jaunts.

Farewell to Dublin

In June 1955, Schrödinger gave a talk in Pisa to a meeting of the Italian Physical Society. Anny accompanied him, and they spent a week touring in Tuscany before heading up into the Alps to escape the summer heat. From Innsbruck, where they collected a new Fiat 2000, they drove to the Tyrol, where they received formal confirmation of Erwin's appointment to the Vienna post in July, while staying at Neustift. The appointment was to take effect from 1 January 1956—with, crucially, full pension rights, even though Schrödinger was only a couple of years short of retirement age—but he was not required to take up residence in Vienna until later in the year.

With all his financial worries resolved, Schrödinger was able to enjoy the rest of what turned out to be his last real holiday in reasonably good health, with two weeks at Lake Garda followed by a visit to Alpbach, taking in the sights of the mountains that he and Anny loved. Then it was back to Dublin, to prepare for the last big move of their lives. It was not a happy return, and things did not improve through the autumn and winter. Almost immediately, Anny admitted herself to her usual clinic for ten days to receive more treatment, and although Ruth March came to help with the preparations for the move, Erwin was struck down by an attack of phlebitis, which prevented him travelling to Cambridge in January to give a planned series of lectures. Although he recovered, Anny needed more treatment in the new year and by the time she was home again Ruth had returned to Innsbruck—just as Erwin was hit with a severe attack of bronchitis. On 11 February, a Sunday, he washed down a cocktail of sleeping pills with a large quantity of whisky. It is unlikely that this was an accident. When Anny had trouble waking him on the Monday morning, she called their doctor, under whose supervision he recovered without having to go to hospital, although he was not allowed to get up until the following Saturday.

Schrödinger's recovery from all this was hardly helped by the ensuing round of parties and formal farewells to his colleagues and other luminaries in Dublin, including Dev and the Irish President, and it was an exhausted man and his wife who finally took the ferry across the Irish Sea on 23 March 1956 on the first leg of their journey, via London, where they stayed for two nights, to Innsbruck, where they arrived on the twenty-eighth. Ruth had her hands full at home, where Arthur March was seriously ill and Hilde was too distracted to be much help, so they didn't linger but drove on to Vienna, where the round of welcomes was as exhausting as, if more cheerful than, the round of farewells had been.

Home is the hero

The return of Austria's greatest scientist to his native land was doubly significant: not only front-page news in its own right, but a sign of the belated normalization of Austrian society, a full ten years after the Second World War had ended. Schrödinger gave his inaugural professorial lecture, on "The Crisis of the Atomic Concept," to a packed audience at the University of Vienna on 13 April 1956. The theme was his familiar message about the nature of reality and the superiority of the wave model to the wave-particle duality of the Copenhagen Interpretation—but he could have been reading from the telephone directory and still have received an ovation from the audience of scientists, civic dignitaries, and old friends eager to welcome him home. Another, more personal, cause for celebration was the marriage of Ruth March, in May 1956, to Arnulf Braunizer; this was soon followed by the news that she was expecting a baby, Erwin's first grandchild, due in February 1957.

A real home for Erwin and Anny proved difficult to find, though, and while the festivities continued the Schrödingers were unable to put down roots: at first they had to live in the Atlanta Hotel, conveniently situated for the Physics Institute but, in Anny's eyes at least, wildly extravagant at a cost equivalent to £2.50 per day for the pair of them on half-board. Eventually, they were able to settle down in a five-room apartment on the third floor of a block on Pasteurgasse, a kilometre or so from the institute. It cost them just over £2,000, but as well as being conveniently located the building had a lift, by now a key requirement for Erwin. The apartment needed refurbishing before they could occupy it, so the move was delayed until after their now regular summer holiday in Alpbach.

Schrödinger's teaching duties were light. He gave lectures twice a week, on "General Relativity" and on "Expanding Universes," but the classes were small and informal; as for the weekly seminar, it was "more like a higher kindergarten, not like the Dublin Seminar."[1] Vienna, however, had delights that Dublin could not compete with, not least the theatre, which the Schrödingers attended frequently, and the proximity of the mountains. But it is debatable whether the "mountain air" really was good for Erwin's health, or whether high altitude was on the contrary detrimental for a man with heart and lung problems. Although he enjoyed an extended summer holiday in Alpbach with Anny that year, heading for the hills as soon as the term was over, in September Erwin was so ill that he had to cancel his planned lectures in Cambridge, already postponed from January. Instead, these talks, known as the Tarner Lectures, were presented by the Professor of Philosophy, John Wisdom, reading from Schrödinger's text. The lectures were published by Cambridge University Press in 1958, under the title *Mind and Matter*; this book was later combined with *What Is Life?* in a single volume, first published in 1967 and still in print.

Mind and Matter is significant as an example of Schrödinger's thinking towards the end of his life, but is otherwise unremarkable. He addresses the old question of whether an objective reality exists if it is not being observed, which leads him into a discussion of whether the world would exist if it were not being observed by "higher animals," asking: "Would it otherwise have remained a play before empty benches, not existing for anybody, thus quite properly speaking not existing?" His answer is that such a conclusion is nonsense, and that consciousness is somehow associated with the process of learning—including the "learning" about its environment involved in the evolution of a plant (or a microbe) to fit its ecological niche. This leads him into murky waters: "We find the behaviour of the individuals of a species having a very significant influence on the trend of evolution, and thus feigning a sort of sham-Lamarckism."

Schrödinger is concerned that this process could adversely affect the evolution of our own species:

> Now I believe that the increasing mechanization and "stupidization" of most manufacturing processes involves the serious danger of a general degeneration of our organ of intelligence. The more the chances in life of the clever and of the unresponsive worker are equalled out by the repression of handicraft and the spreading of tedious and boring work on the assembly line, the more will a good brain, clever hands and a sharp eye become superfluous. Indeed the unintelligent man, who naturally finds it easier to submit to the boring toil, will be favoured; he is likely to find it easier to thrive, to settle down and to beget offspring.

> The result may easily amount even to a negative selection as regards talents and gifts. . . .
>
> Instead of letting the ingenious machinery we have invented produce an increasing amount of superfluous luxury, we must plan to develop it so that it takes off human beings all the unintelligent, mechanical, "machine-like" handling. The machine must take over the toil for which man is too good, not man the work for which the machine is too expensive.

Well, maybe Schrödinger's conclusion was right, although possibly not reached for the right reason!

Elaborating his discussion of the nature of reality, Schrödinger then bemoans the fact that in physical science, since the time of the ancient Greeks, "a moderately satisfying picture of the world has only been reached at the high price of taking ourselves out of the picture, stepping back into the role of a non-concerned observer." Jung, he says, was right when he pointed out that all science is "a function of the soul, in which all knowledge is rooted. The soul is the greatest of all cosmic miracles, it is the *conditio sine qua non* of the world as an object." Schrödinger's own conclusion is that "subject and object are only one. The barrier between them cannot be said to have broken down as a result of recent experiences in the physical sciences, for this barrier does not exist."

The consciousness that Schrödinger refers to is, as ever, not an individual consciousness, but the collective consciousness of which we are all a part. And his final lecture addresses issues of striking personal relevance—religion and life after death. I won't spell out his argument, which is based on a statistical interpretation of time, but the conclusion, no doubt appealing to someone in his seventieth year, was that "we may, or so I believe, assert that physical theory in its present stage strongly suggests the indestructibility of Mind by Time."

Declining years

Throughout most of his adult life, one of Schrödinger's driving motivations had been the need to secure his and Anny's—particularly Anny's—financial future. He had at last achieved it, but at the age of sixty-nine had little else. "The only thing I have enough of now," he wrote, "is money."[2]

The reason for this gloomy comment was essentially his declining health. The cold and damp of winter in Vienna exacerbated his heart and lung problems, and with the blood supply to his brain restricted, after giving lectures he became very tired, and on several occasions confused and rambling—that is, when he could give his lectures at all. Although Anny always drove him the short distance from their apartment to the institute and back again, in the winter term he was able to teach fewer than half his classes. It was in the apartment that Schrödinger held court, when he was well enough, receiving many friends from the artistic community as well as scientists. Even here, after November 1956, when Soviet tanks crushed the uprising in Hungary, the atmosphere was marred by the widespread fear that Austria, so recently freed from occupation, might be the next in line.

December also brought gloomy news, but at a more personal level. On a visit to Vienna, Ruth, now seven months pregnant, brought word that Arthur March was terminally ill, with cancer of the throat. Ruth and her new husband were looking after things for Hilde in Innsbruck; the Schrödingers gave them some money to reduce their worries at this difficult time.

Having "enough" money also meant that in spite of everything Anny could enjoy the Christmas festivities. She wrote to her friend Elisabeth Ullmann:

> What an experience when I came on December 29 to Fritz the Baker's. When one has lived twenty years in the British Isles one thinks such bounty is impossible, a little shop with about 20 serving girls and the Chef whom they treat like a lord[. It is] quite expensive and every variety of baked goods can be found, croissants, rolls, salt rolls, kümmel buns, sour breads, sandwich bread and different kinds of black breads, brioches, milk sticks, pretzels and bosniaks, not to forget six different kinds of tarts.

Clearly, Anny liked her baked goods, and no doubt some of these found their way on to the table when the Schrödingers entertained a large party of friends on 31 December; but the festivities were short-lived for Erwin, who went down with another attack of bronchitis as the old year gave way to the new. Although severely ill, he was saved with the help of aureomycin, the first of the tetracyclines, penicillin-like drugs developed after the Second World War. So he was as well as he could now expect to be when Ruth arrived in February to stay with the Schrödingers while awaiting the arrival of her baby; she had been sent away from the March household by Arnulf, to get her away from the dying Arthur March and the distraught Hilde while he held the fort there.

Erwin's grandson, Andreas Braunizer, was born at the University of Vienna Clinic on 28 February 1957. A few weeks later, on 17 April, Arthur March died. In a letter to Hilde, Erwin wrote of his belief that a dying person should be allowed to slip away peacefully, aided by doses of opium, rather than suffer the indignity of prolonging life for a few extra days just for the sake of it. He was clearly contemplating his own fate, but told her: "I am glad of the few years [left], for the world is very, very beautiful."

Those few years almost became a few weeks. In May, Anny was being treated for asthma at a clinic near Innsbruck when she was summoned back to Erwin's bedside: he was close to death with a severe bout of pneumonia, and was saved only by a cocktail of the new antibiotics, including penicillin, streptomycin, Terramycin, and Magnamycin. Ten years earlier he would surely have died; as it was, by the end of the month he had turned the corner. There was double cause for celebration on 31 May—for the same day Germany's top civil order, the "Pour le Mérite,"[3] was formally conferred on Schrödinger at a ceremony in Bonn, although of course he had known about the award for some time. His friend Lise Meitner received the decoration at the same time—only the second woman to be honoured in this way.

Because of his poor health, Schrödinger was formally excused any teaching for his last year as a professor in Vienna, 1957–58. But he was well enough to visit the Physics Institute from time to time, and to participate in a joint meeting of the Austrian Physical Society and the Chemical–Physical Society in March 1958. It was there, on 26 March, that he gave his last scientific lecture, at the age of seventy. The theme went back half a lifetime to focus on the links between energy and entropy, and to argue that the law of conservation of energy is true only in a statistical sense.

But entropy, statistical or not, was fast catching up with Schrödinger. In the spring of 1958, he suffered an attack of phlebitis which required hospitalization and then lengthy bed rest at home; he recovered just in time for the regular summer visit to Alpbach. By now formally retired, on 30 September 1958, a month after his seventy-first birthday, Schrödinger was appointed to the honorary post of professor emeritus at the University of Vienna. He would have just over two years to enjoy the secure old age he had striven so hard to achieve.

The triumph of entropy

The first of those years was everything Schrödinger had hoped it might be. After enjoying the role of Grand Old Man in Vienna, in the summer he took off as usual for the Tyrol with Anny; he stayed there for four months, at times enjoying the company of old friends, including Lise Meitner. In October the Schrödingers stayed in the Italian city of Bolzano, the largest city of the South Tyrol (and now the home of the ice mummy "Ötzi"), where their delight at the beautiful mountain scenery was offset by a recurrence of Erwin's bronchitis as the autumn faded into winter. One side-effect of this bout of ill-health was that, often unable to sleep, he spent many long night hours writing letters, one of which in particular provides an insight into his thoughts at this time about quantum mechanics.

The letter was sent to John Synge (1897–1995), a distinguished Irish mathematician and physicist who had joined the Dublin Institute for Advanced Studies as a senior professor in 1948, and was one of its leading lights for the rest of his career. Schrödinger's letter castigated physicists in general for their (as he saw it) unthinking acceptance of the Copenhagen Interpretation: "With a very few exceptions (such as Einstein and Laue) all the rest of the theoretical physicists were unadulterated asses and I was the only sane person left."

He complained that his efforts to get people to take the puzzle of wave-particle duality seriously had fallen on stony ground because his colleagues "have formed the opinion that I am—naturally enough—in love with 'my' great success in life [and] therefore, so they say, I insist upon the view that 'all is waves.'" In their view,

> old age dotage closes my eyes towards the marvelous discovery of "complementarity." So unable is the good average theoretical physicist to believe that any sound person could refuse to accept the Kopenhagen oracle [Niels Bohr]. . . . If I were not thoroughly convinced that the man is honest and really believes in the relevance of his [idea] I should call it intellectually wicked.

Nothing could be clearer, and Schrödinger would surely be delighted to see the way the Copenhagen Interpretation has fallen from favour today, although he might not approve of all the ideas that are offered as alternatives.

The letter to Synge was just about Schrödinger's last word on quantum mechanics. As the mountains grew colder, Erwin and Anny moved down into Italy for three weeks, stopping at Mantova, Cremona, Piacenza, Parma, Verona, and Venice before returning to Vienna to hole up for the winter. Schrödinger stoically put up with the respiratory problems that he had become used to, but by the spring of 1960 it was clear that the problem this time was more than just bronchitis, and tests revealed a recurrence of the tuberculosis that he had suffered in the early 1920s. At least there were now effective drugs to tackle the problem, but part of the treatment still involved the old idea of fresh air and sunshine. So

Schrödinger was packed off to Alpbach to stay. There, he took the opportunity to finish the second part of what became his final book, which would be published in German as *Meine Weltansicht* and in English as *My View of the World*.

The first part of the book was essentially the essay on metaphysics, the Vedanta, and consciousness written in 1925, shortly before Schrödinger succeeded Planck as professor in Berlin. Nor was the "new" material really new, being more a restatement of Schrödinger's ideas about the relationship between mind and matter, addressing the question "What is real?" and endorsing the belief that "we living beings all belong to one another, that we are all actually members or aspects of a single being, which we may in western terminology call God, while in the Upanishads it is called Brahman." Schrödinger says that he has "no hesitation in declaring quite bluntly that the acceptance of a really existing material world, as the explanation of the fact that we all find in the end that we are empirically in the same environment, is mystical and metaphysical." Although anyone who wishes to take that view is at liberty to do so, "he certainly does not have the right to pillory other positions as metaphysical or mystical on the supposition that his own is free from such 'weakness.'"

During his "cure" at Alpbach, Schrödinger also had plenty of time to write letters to his wide circle of friends and acquaintances. In terms of his long battle with what might be regarded as the quantum-mechanical establishment, the most significant of these was a letter to his old sparring partner Max Born. "I do," he said,

> need to give you once a thorough head washing. . . . The impudence with which you assert time and again that the

> Copenhagen Interpretation is practically universally accepted, assert it without reservations, even before an audience of the laity—who are completely at your mercy—it's at the limit of the estimable. . . . Are you so convinced that the human race will succumb before long to your own folly?

As someone who was taught the gospel of the Copenhagen Interpretation just a few years later, and only much later still appreciated the nature of its folly, I find these words from Schrödinger in 1960 strike close to my heart!

By the beginning of November, Erwin had had enough of Alpbach, and decided to go back to Vienna; Anny had been taken seriously ill with asthma and rushed to hospital on 20 October. He returned to their apartment on 9 November, driven in the Schrödinger car by a friend. In truth, he was not well enough to look after himself, and struggled along, with help from Anny's sister and the couple who looked after the apartment building, only until 2 December, when he was admitted to hospital. Just a few days later, Anny was well enough to go home—she would live for another four years—and Erwin now demanded to return home himself, saying, "I was born at home and I'll die at home." His wish was not immediately granted, and as his condition deteriorated over the next few weeks he remained in hospital, with Anny at his bedside, holding his hand. But he was at last returned to his own bed on the morning of 3 January 1961. He died the following day, at 6:55 p.m. The cause of death was simply given as old age; he was seventy-three.

Schrödinger had asked to be buried in the little churchyard at Alpbach. The priest at first demurred, on the not unreasonable grounds that Erwin was not a Catholic; but when he was advised that Schrödinger had been a member of the Pontifical Academy he relented, and the interment took place on Sunday, 10 January. Schrödinger was gone, but his scientific legacy lived on.

Chapter Fourteen

Schrödinger's Scientific Legacy

The most significant development in quantum physics since 1960—arguably the most significant development in science in the twentieth century—was the resolution of the EPR "Paradox" and experimental confirmation that quantum entanglement (a term coined by Schrödinger) is real. Apart from its implications for our understanding of the Universe we live in, this has led to practical benefits, including huge economic and commercial benefits, that are just beginning to be realized. The story begins with an American Communist who shared Schrödinger's dislike of the Copenhagen Interpretation, develops through the work of an Irish mathematician who knew better than to believe everything he read in books, and reaches a climax (although not an end) in the laboratory of a Frenchman who risked his career by attempting to prove what others regarded as impossible.

Hidden reality and a mathematician's mistake

David Bohm (1917–92) was an American physicist who wrote a classic book presenting the Copenhagen Interpretation, but as a result of his careful analysis of this view decided that it was nonsense, and developed an alternative understanding of the quantum world. His book, *Quantum Theory*, was published in 1951, when Bohm was under a cloud following an investigation by the notorious Un-American Activities Committee. He had briefly been a member of the Communist Party, in 1942 (when the USA and USSR were allies in the Second World War), and although he was cleared of any "un-American" behaviour, such was the atmosphere of paranoia at the time that he had been dismissed from his post at Princeton University and was hounded into exile not long after the book was published. He worked for a time in Brazil, then in Israel, and at the University of Bristol, in England, before settling, in 1961, at Birkbeck College of the University of London.

Bohm's unease with the Copenhagen Interpretation stemmed from discussions with Einstein at Princeton following the publication of his book. Even before he left the USA, and in spite of the turmoil in his private life, Bohm had written two papers, published early in 1952, that outlined his alternative understanding of quantum mechanics. This was essentially a more thoroughly worked-out version of the pilot wave idea that had been suggested by Louis de Broglie back in 1927, but had been unjustifiably ignored because of the way Bohr and his colleagues had steamrollered all opposition to the Copenhagen Interpretation. Indeed, the steamrollering had been so effective that Bohm was unaware of de Broglie's pilot wave model when he worked out his own variation on the theme. In essence, the idea is that entities such as electrons are real particles, which are guided by a wave which obeys the Schrödinger equation. Very crudely speaking, as I have mentioned, an analogy can be made with a surfer riding a wave on the ocean.

The snag with this idea is that the pilot wave has to "know" about everything likely to affect the trajectories of the particles (in principle, everything in the Universe) in order to guide each particle to its destination. It is said to be influenced by so-called "hidden variables." If we knew what the hidden variables were, we could use them to calculate the quantum behaviour of electrons and other particles without resorting to the collapse of the wave function or the statistical interpretation. As the Oxford physicist David Deutsch (b. 1953) has put it, "a non-local hidden variable theory means, in ordinary language, a theory in which influences propagate across space and time without passing through the space in between: [in other words] they propagate instantaneously."[1]

Apart from the momentum of the Copenhagen juggernaut, there was another reason why most physicists did not take hidden variables theory seriously in the 1950s. In 1932, John von Neumann (1903–57), a Hungarian-born mathematical genius, had published a book in which, among other things, he "proved" that hidden variables theories could not work. His contemporaries were so in awe of von Neumann's ability that for a generation this proof was barely questioned, and it was widely cited as gospel, without being spelled out in full, in standard texts such as Max Born's *Natural Philosophy of Cause and Chance*, published in 1949. Born praised von Neumann's "brilliant book" which proved that "no concealed parameters can be introduced with the help of which the indeterministic description could be transformed into a deterministic one." One of the people who was deeply impressed by this when he read Born's book, just after it was published, was a young man in his final year as a student in Belfast—John Bell (1928–90). But since Bell did not read German, and von Neumann's book had not been published in English, he had to take Born's word for it that von Neumann knew what he was talking about.

Bell, who had been born in Belfast, came from a "poor but honest" family.[2] With an older daughter and two younger sons besides John, his parents could barely afford for one child to go to secondary school, but John's ability was so obvious at an early age that they determined to give him every opportunity they could. Their struggle was rewarded when John qualified to study at Queen's University at the age of sixteen, a year before he was old enough to enrol there. After working as a lab assistant while waiting to start his course, he went on to obtain a degree in experimental physics in 1948 and another degree in mathematics in 1949, the year he read Born's book. When he had first encountered quantum mechanics as an undergraduate, he had been outraged. "I did not dare to think it was false," he later told Jeremy Bernstein, "but I knew it [the Copenhagen Interpretation] was rotten."[3]

After graduating, Bell worked for the UK Atomic Energy Research Establishment until 1960, when he moved to the European particle physics research centre CERN, in Geneva, where he stayed for the rest of his career. He had read Bohm's papers on hidden variables theory and been impressed by them, so he was already having doubts about von Neumann's "proof." He later wrote: "In 1952 I saw the impossible done. It was in the papers by David Bohm. Bohm showed explicitly how parameters could indeed be introduced into non-relativistic wave mechanics, with the help of which the indeterministic description could be transformed into a deterministic one."[4]

When he got a German-speaking colleague to translate the relevant part of von Neumann's book for him, he saw at once where von Neumann had gone wrong. But Bell did not have an opportunity to look deeply into the implications until 1963–64, when, having taken a sabbatical year off from his work at CERN, he visited the Stanford Linear Accelerator Center (SLAC) and other academic centres in the United States. Von Neumann had perpetrated a howler. His "proof" didn't work because he had made a ridiculous assumption, equivalent to saying that if the average height of a group of children is 1.2 metres then each child has a height of 1.2 metres. If that sounds silly to you, you are in good company: Bell thought so as well. In 1988 he said: "The von Neumann proof, if you actually come to grips with it, falls apart in your hands! There is nothing to it. It's not just flawed, it's silly! . . . When you translate [it] into terms of physical disposition, they're nonsense. You may quote me on that: The proof of von Neumann is not merely false but foolish!"[5]

But the fact that von Neumann was wrong didn't necessarily mean that Bohm was right. Bell decided to go back to the roots of the non-locality debate, starting out from the EPR thought experiment, to see whether or not non-locality was a fundamental feature of all hidden variables theories. This was the first step towards the most profound change in our understanding of the quantum world since the revolution of 1926–27.

The Bell test and the Aspect experiment

Bell started out from a version of the EPR experiment devised by Bohm and his colleague Yakir Aharonov (b. 1932) while Bohm was working in Israel in the late 1950s. In this version of the experiment, instead of thinking in terms of momentum and position the imaginary experiment deals with the property known as spin—or the equivalent property of photons, polarization. Simplifying the argument somewhat, there are circumstances in which quantum processes will produce two electrons which fly off in different directions, and which have a combined spin of zero. So if one has spin up, the other has spin down, and so on. There is an extra degree of variability because the spin can be measured in any direction (for example, horizontally or vertically, or any angle in between). If the spin of one electron is measured in any particular orientation, then we know that the spin of the other electron must be the opposite, for that particular orientation.

But how does the other electron know this? The naive answer is that the electrons set out on their respective journeys with well-defined but opposite spin states. But a more subtle experiment removes this possibility. If the spin of one electron is measured for one orientation, and the spin of the other electron is measured for a different orientation, the situation is not so simple. The relationship between the measured spins can be predicted from quantum mechanics in a well-defined statistical way, but the predictions do not match those of "common sense." When Bell calculated the appropriate numbers for the correlation, he found a difference between the prediction of quantum mechanics and the predictions of any theory based on local reality and hidden variables. He expressed this in terms of a relation called Bell's inequality, which says that if local reality holds, one set of numbers determined by experiment must be smaller than another set of numbers determined by experiment. This is a bit like saying that in the everyday world the number of red-haired men must be less than the total number of men and women. But if quantum mechanics is correct, and local reality does not hold, then Bell's inequality is violated, as if there were more red-haired men in the world than all of the men and women put together.

"I got that equation into my head and out on to paper within about one weekend," he later told Paul Davies. "But in the previous weeks I had been thinking intensely all around these questions. And in the previous years it had been at the back of my head continually."

Bell was well aware that, unlike the situation originally described by Einstein, Podolsky, and Rosen, his hypothesis could actually be tested by experiment, taking the EPR debate out of the realms of philosophy and in to the lab. In the paper he wrote describing these results he said: "It requires little imagination to envisage the measurements involved actually being made." But few people noticed the paper, or this extraordinary claim, because it appeared in an obscure journal.

Such an important theoretical discovery should have been published in the prestigious *Physical Review*. But *Phys Rev*, as it is known, charged for publication, on the basis of the number of pages printed. Although Bell's paper was just six pages long, as a visitor to Brandeis University (where he had arrived in the course of his sabbatical trip) he felt diffident about asking the institute to pay, and sent it to the journal *Physics*, which didn't levy page charges. Unfortunately, it didn't have many readers, either. So it was five years before Bell received a letter from John Clauser (b. 1942), a Berkeley-based physicist who had read the paper and intended to carry out an experiment to test whether or not Bell's inequality was violated. Bell enthusiastically replied that:

> In view of the general success of quantum mechanics, it is very hard for me to doubt the outcome of such experiments. However, I would prefer these experiments, in which the crucial concepts are very directly tested, to have been done and the results on record. Moreover, there is always the slim chance of an unexpected result, which would shake the world![6]

But it was 1972 before Clauser and his colleagues reported that their experiment (actually using photons rather than electrons) had indeed produced results in line with the predictions of quantum theory, and against local reality.

Exciting though these conclusions were, the experiments were not definitive. They involved using beams of photons, rather than individual pairs, and the detectors used could detect only a fraction of the photons in the beams. Conceivably, all the undetected photons were behaving in a different manner that would lead to different conclusions. Clauser's experiment was a pointer to the way ahead, rather than the last word on the subject. But the fact that the experiment had been carried out at all was a triumph, which stimulated other researchers to attempt improvements on the theme. The one who achieved the most striking success was a Frenchman, Alain Aspect (b. 1947), who spent three years doing voluntary work (a version of French national service) as a teacher in Cameroon after obtaining his degree, and read up on quantum physics, including the EPR Paradox and its implications, during his spare time in Africa. When he returned to France in 1974, he was determined to make an experimental test of the Bell inequality for his doctoral work at the University of Paris–South. When he visited Bell in Geneva to discuss his plans, Bell's first question was "Do you have tenure?," meaning "Do you have a secure permanent post?" When Aspect replied that, far from having tenure, he was still only a PhD student, Bell replied: "You must be very courageous."[7] He meant that if Aspect failed in his quest to carry out the difficult experiment he might end up with no PhD and no prospect of a career in physics. But although it took longer than he had anticipated, Aspect eventually achieved his goal.

The important feature of all such experiments is that measurements made of photon beam *A* produce a set of random numbers. The measurements made of photon beam *B* also produce a set of random numbers, so no information about photon beam *A* can be gleaned just by looking at those numbers. But when the two sets of numbers are put together and compared, it is possible to see a correlation between them—to see that the "answers" obtained from photon beam *B* depend on the "questions" asked, simultaneously, of photon beam *A*, no matter how far away photon beam *A* is. The only way to make this comparison, though, is to convey information from *A* to *B* by conventional means (meaning, more slowly than the speed of light), so there is no conflict with relativity theory.

In the first versions of this kind of experiment, there was a possible loophole. The experiments were set up so that the entangled photons went their separate ways and passed through two different sets of detectors. But the whole setup was fixed in advance, so it was possible to argue that this determined the outcome of the experiments, with no need for any "spooky action at a distance" linking the parts of the experiment. As well as greatly improving the experimental statistics, Aspect closed this loophole by devising a system which changed the setting of the polarizing filter used to measure the photon beam at random while the photons were in flight through the experiment. The two sets of measuring equipment were 13 metres apart, so the switching had to take place in less time than it takes for a message travelling at the speed of light to cross 13 metres, which is about 43 nanoseconds (1 nanosecond is 1 thousand-millionth of a second). In round numbers, the switching between settings actually took place once every 10 nanoseconds (every 10 thousand-millionths of a second). In that way, there was no possibility that the setup at detector *B* could "know" what the setting of detector *A* was at the moment photon *B* arrived. Aspect himself said:

> What these experiments have shown is first that they violate Bell's inequalities, and on the other hand that these results are in very good agreement with the predictions of quantum mechanics. So we assume that quantum mechanics is still a very good theory. [But] even in this kind of experiment it is not possible to send any messages or useful information faster than light.[8]

Every subsequent refinement of this kind of experiment has confirmed the accuracy of this statement.

The results of Aspect's experiments were published in 1981 and 1982, and he received his PhD in 1983. It was this definitive confirmation of the validity of quantum mechanics, closing the last major loophole for local realistic theories, that prompted me to write my book *In Search of Schrödinger's Cat*, which appeared in 1984, offering a historical account of the development of quantum physics. I was (and still am) happy to accept the evidence that "it is not possible to send any messages or useful information faster than light." But a few maverick physicists made a series of wrong-headed but sophisticated attempts to prove that entanglement offers a way to send messages faster than light. They were ultimately proved wrong, but research partly stimulated by their quest has led quantum physics into the almost equally exotic realms of cryptography and teleportation.

Quantum cryptography and the "no cloning" theorem

One of the people who tried to devise a way to use entanglement to send messages faster than light was Nick Herbert, an American physicist who had completed a PhD at Stanford University in 1967, but had been unable to get an academic post and had worked in a variety of jobs while pursuing his interest in quantum physics in his spare time, together with a group of like-minded thinkers based in Berkeley, California. The details of Herbert's design for faster-than-light (FTL) signalling do not matter, except for one key point. It depended on making exact copies of photons—"clones," in the vernacular. A copy of Herbert's paper found its way to Wojciech Zurek (b. 1951), a Polish-born American physicist who found the flaw in the argument. Together with William

Wooters he proved that "a single photon cannot be cloned," and used that phrase as the title for a paper published in *Nature* in October 1982 which began: "Note that if photons could be cloned, a plausible argument could be made for the possibility of faster-than-light communication." Although it ruled out FTL signalling, it was the "no cloning theorem," also discovered independently by the Dutch physicist Dennis Dieks, that opened the way to the practical use of quantum entanglement to create uncrackable codes—quantum cryptography.

There are several approaches to the problem of quantum cryptography, but they all depend on code systems that use a "key" of random numbers. The description here is adapted from my book *Schrödinger's Kittens*, since I could not see any way to improve on it.

The kind of code I want to describe is familiar from spy stories. The two people using the code (always referred to by cryptologists as "Alice" and "Bob") are each equipped with an identical list of random numbers, a so-called "pad," which can be as thick as a telephone book. The person sending the message turns it into numbers (perhaps as simply as by allotting the number 1 to letter *A*, 2 to *B*, and so on), and then chooses one of the pages of random numbers from the pad. The numbers on the pad are then written out under the numbers corresponding to the letters in the message, and the pairs of numbers are added together. The coded message is then sent, along with information about which page on the pad was used, and at the other end the same set of random numbers are subtracted from the coded signal to restore the original message. The code is called the Vernam cipher, after the American Gilbert Vernam, who developed it during the

First World War, and is sometimes referred to as the "one-time pad" technique, because spies were supplied with the random number book in the form of a tear-off pad, so that each sheet could be used once and then destroyed (if the same set of random numbers from the same page of the pad is used to code more than one message, patterns recur which make it possible to crack the code).

This kind of code cannot be broken, unless the person who intercepts it also has a copy of the same one-time pad. The snag is that under the conditions in which espionage agents usually operate it is all too likely that the interested third party will get hold of a pad; worse, it is possible for the third party to have a copy of the pad, and to be breaking the code, without the two users of the code knowing.

Quantum physics offers a way around both problems. There is no need to keep the coded messages themselves secret, because they are useless without the information that comes over a quantum channel—the key. What is needed is a way to communicate the key itself—a string of random numbers—from Alice to Bob in an uncrackable way. To make things as simple as possible, the string of numbers can be in binary arithmetic, a string of 0s and 1s like the code used by computers; so the key can be transmitted as any system of on/off, either/or signals.

Charles Bennett, of the IBM Research Center at Yorktown Heights in New York, and his colleagues showed that you can do this using polarized light. The technique involves Alice sending Bob a stream of photons, polarized either up or across one of two agreed orientations (at 45 degrees to each other), but with the polarization of each photon chosen at random. Bob measures the polarization of the incoming photons, but for each measurement he can only align his detector with one of the agreed directions of polarization—again, chosen at random. In each case, he will get an "answer' corresponding either to vertical (binary 1) or horizontal (binary 0) polarization relative to his detector. He then tells Alice which orientations he used for each measurement, and she tells him which of these match the way the photons were sent (this communication can be by email). Bob and Alice then throw out all the measurements for which Bob chose the "wrong" polarization, and are left with a string of 1s and 0s, their secure key in binary code. It sounds like a tedious process when spelled out like this, but in the real world anyone using such a system would avoid the tedium by running it through a computer which did the donkeywork.

The great beauty of the technique is that the only way a third party can find out what code Alice and Bob are using is to "eavesdrop" on the quantum communication channel, and measure the polarization of the photons as they pass through. But the act of measuring the polarization of a photon changes the polarization! The photon cannot be copied (cloned) and left unchanged. Even if the eavesdropper copies the measured photon and sends it on to Bob, it will have been randomized. Bob and Alice can check for this interference by standard techniques which, in effect, compare every fifth, or every seventh, or whatever, letter of the key, without revealing the whole key.

Artur Ekert (b. 1961), a Polish physicist now at the University of Oxford (who has also collaborated with Bennett), found another way to achieve the same ends, showing how the required random string of digits could be obtained from a variation on the EPR experiment itself. The

EPR photons are fired off in opposite directions, suitably entangled but as yet unmeasured, one beam to Alice and one to Bob. Alice and Bob can each measure the polarization of their photons, using detectors oriented at random along one of a set of previously agreed polarization directions. They then tell each other, over any ordinary public communications channel, which measurements they made, but not the results of those measurements. Finally, they discard the measurements where they used different orientations, and construct their secure key out of the results of the measurements where their polarization detectors were aligned the same way—making allowance, of course, for the fact that the members of each pair of EPR photons have the opposite polarization, once they have been measured, so that Bob always gets a 1 where Alice gets a 0, and vice versa. And, once again, any attempt to "tap" the quantum communication channel by looking at the photons before they get to Bob and Alice will scramble up their polarizations in a detectable way. It's worth mentioning Ekert's variation on the theme if only because it led to a memorable encounter between him and John Bell, when Bell was on a visit to Oxford and Ekert was just a graduate student. After a talk Bell gave, Ekert was able to meet him briefly and explain his idea. Bell was astonished. "Are you telling me that this could be of practical use?" he asked. Ekert said it could. "Well," said Bell, "it's unbelievable."[9]

It may all sound far-fetched and improbable; but even by the time *Schrödinger's Kittens* was published, in 1995, Bennett and his colleagues had actually built a system which works in this way. Admittedly, in the prototype the coded messages were only sent across a distance of 30 centimetres; but that is because they built it on a desk-top. I wrote in 1995 that "in principle, polarised photons could be sent unaltered through several kilometres of optical fibre. And after all, when John Logie Baird built the first television transmitter it only sent a picture over a distance of a couple of metres."

The quantum cryptographers have since more than met my expectations. In 2004, appropriately enough in Schrödinger's home city of Vienna, a team of physicists headed by Anton Zeilinger (b. 1945) carried out the first electronic bank transfer of money using quantum cryptography to ensure the security of the data. This wasn't just a casual experiment involving one physicist transferring a small amount into another physicist's account but a full-blown transfer of official funds between a major bank and the mayor's office. In 2007, during a Swiss election, quantum encryption was used to ensure the security of votes cast electronically in Geneva. Before long, this is likely to be the way the security of information routinely transmitted over the Internet is assured. And it need not just be information that is transmitted.

Quantum teleportation and classical information

Quantum teleportation also uses the "no cloning" theorem, but in a slightly different way. A photon—or other quantum entity—cannot be duplicated (cloned), but all of its properties can be transferred to a second photon even though (indeed, because) the first photon is changed. In effect, the second photon has become the first photon.

Entanglement and action at a distance are at the heart of the technique of quantum teleportation, which was also proposed by Charles Bennett, and published in the journal *Physical Review Letters* in 1993.

In the classical, everyday world, sending copies of things to distant places is routine. The obvious analogy with teleportation is the fax machine, which has the added advantage of leaving the original copy intact at its starting place while producing a duplicate at the destination. Newspapers and books are reproduced in editions containing hundreds of thousands of essentially identical copies, as far as their information content is concerned.

But at the quantum level, copying runs into difficulties. The first is simply a question of detail. The uncertainty principle makes it impossible to know every detail about every atom in, say, a sheet of paper, or even the exact position of every molecule of ink in the printing on the paper; so the faxed "copy" can only ever be an approximation. In addition, scanning an object at the quantum level changes its quantum state. But this apparent problem is what makes quantum teleportation possible! Even if you did obtain the information needed to build a copy of a quantum system, the original would be destroyed. This is more like the science-fiction version of teleportation ("Beam me up, Scotty") than the way a fax machine works.

Classical information can be copied, but can only be transmitted at the speed of light (or less); quantum information cannot be copied, but sometimes, as in the EPR experiment, it seems to propagate instantly from one place to another. Bennett and his colleagues used a mixture of these classical and quantum features of a system to devise their teleportation device.

They describe this in terms of two people—again, Alice and Bob—who want to teleport an object. In this teleportation for beginners, the object to be teleported is simply a single particle—perhaps an electron—in a particular quantum state. At the beginning of the experiment, Alice and Bob are each given a box containing one member of a pair of entangled objects, equivalent to their each carrying one of the photons from the EPR experiment, without measuring its polarization. Then, they go off on their travels across the Universe. Some time later—perhaps many years later—Alice wants to send another particle to Bob. All she has to do is to allow the "new" particle to interact with her entangled particle, and to measure the outcome of their interaction. This both establishes and changes the state of her entangled particle, and instantaneously establishes and changes the state of Bob's entangled particle in an equivalent way.

Bob doesn't know this yet, because he is somewhere on the other side of the Universe. So now Alice has to send him a message, perhaps by radio, or perhaps by putting a notice in the newspaper that Bob reads every day, telling him the result of her measurement.

This message contains only classical information, so she can send as many copies as she likes in as many newspapers or radio broadcasts as she likes. Eventually, Bob will get the message. Armed with the information about how the interaction between Alice's two particles turned out, Bob can now look at his own entangled particle and use the information to "subtract out" the influence of his own original particle from its present state. What he is left with is an exact copy of the other particle—the one that Alice wanted to send to him. And she has done this without knowing where Bob is, or even speaking to him directly. The original version of the third particle was destroyed (changed into another quantum state) when Alice carried out her measurement on it, so Bob's version is unique, unlike a newspaper, and he is fully entitled to regard it as the original particle, conveyed to him by a combination of classical message and action at a distance.

This process, Bennett stresses, defies no physical laws, and only permits teleportation to take place at less than the speed of light—Bob needs Alice's "classical" message in order to untangle his particle properly, and if he looks at his particle too soon he will change its quantum state and destroy any prospect of untangling it in the right way. "Alice's measurement forces the other EPR particle to change in such a way that the classical information that comes out of her measurement enables someone else to produce a perfect copy of what went in," but "it cannot take place instantaneously."[10] It is, as more than one wag has remarked, "teleportation, Jim, but not as we know it." And once again, in the years since 1995 experimenters have put the idea into practice on an impressively large scale, although not yet from one side of the Universe to the other.

The first successful quantum teleportation experiments on the laboratory scale (over about a metre) were carried out in Innsbruck in 1997, by Zeilinger's team. This was soon extended to a range of a few hundred metres, using fibre optics, and by 2004 different teams were carrying out partial teleportation of the states of whole atoms.[11] In May 2010, a team of Chinese scientists reported in *Nature* successful quantum teleportation over a range of 16 kilometres sending laser beams through the air; and there are already plans (but no funding) for a satellite to be launched carrying photons in an entangled state that can be teleported back to Earth from orbit. But all of these ideas pale in comparison with the most important practical application of entanglement, which uses the same technology that is needed for quantum cryptography and quantum teleportation. This is the development of the quantum computer, a practical proposition as "unbelievable," in the sense used by John Bell that day in Oxford, as any development in the application of science to technology.

The quantum computer and the Multiverse

I have described the background to the development of quantum computers in detail in my book *In Search of the Multiverse*, but the mind-blowing implications can be understood without going into the nuts and bolts of the quantum engineering. The essence of modern electronic computers is that they carry out calculations using binary code, which can be represented as a string of 0s and 1s, or in practical terms as an array of switches which can be either on or off. Each "on/off" unit is called a bit. Eight bits make a byte, and the power of a computer is often expressed in terms of the number of bytes in its "brain"—these days, many gigabytes in even a modest laptop or smart phone. The essence of a quantum computer is that each of the switches—each bit—can exist in a superposition of states, without any collapse of the wave function, so that it is both on and off (storing both 1 and 0) at the same time. Crucially, such switches (or "stores") can also be entangled, so that a pair of these so-called qubits may be guaranteed to be either both in the state 1 or both in the state 0, even though neither of the qubits can be said to be definitely in the state 0 or definitely in the state 1. In principle, each qubit could be a single atom or a polarized photon; as yet, in practice molecules containing several atoms are used, and each bit of information is stored in billions of copies of each molecule.

All of this means that the power of a quantum computer is literally exponentially greater than the power of an equivalent conventional computer. A quantum computer using n qubits is 2^n times as powerful as a conventional computer with n bits. For example, with 2 qubits a quantum computer would be equivalent to a conventional computer with 4 bits, with 4 qubits a quantum computer would be equivalent to a conventional computer with 16 bits, and with just 10 qubits a quantum computer would be equivalent to a conventional computer with 1,024 bits, or one kilobit.[12] The crucial point is that modest quantum computers have been built and do work as expected, confirming that they operate in line with the rules of quantum physics, including entanglement.

The most dramatic claims for success in this field have come from a Canadian commercial company, D-Wave Systems, which announced in May 2011 that it had sold a 128-qubit quantum computer to a security firm, Lockheed Martin. Nobody outside the company knows exactly how the computer works, except that it involves linking sixteen 8-qubit units, because details have been kept secret for commercial reasons. This has led to doubts among some scientists as to whether it really operates purely on quantum principles. But it certainly does something—the Lockheed Martin people are no fools, and spent a year reviewing the machine before buying it. And even if it turned out that "only" the 8-qubit units are really operating purely as quantum processors, which there is no reason to doubt since similar mini-processors have been made to work in several labs, that strikes at the heart of our understanding of reality and leads us straight back to Schrödinger's view of the world.

David Deutsch, who is a pioneer of the theory of quantum computation, has asked: Where do the calculations carried out on a quantum computer take place? Remember that there is no collapse of the wave function, exactly like the situation in the Schrödinger's cat thought experiment before the box is opened. So both possibilities are real. If a quantum computer using 8 qubits is equivalent to a conventional computer using 256 bits, that is because there are 256 separate computers corresponding to each possible quantum state of the processor carrying out the computation in 256 different universes—the "parallel worlds" of science fiction. They work on the problem together, solve it, and share the answer. "That's what the laws of physics tell us," says Deutsch. "It can't be that there are multiple universes at the level of atoms but only a single universe at the level of cats."[13]

This does not mean that the Universe somehow "splits" into—in this case—256 copies of itself when you run the computer. The Multiverse idea says that there always were 256 universes, identical to one another up to the point where the computation is run, and that the identical experimenters in each of those universes each decide to carry out the same experiment—hardly surprising, since they are identical. And this "no collapse" picture is very close to what Schrödinger suggested in 1952 (see Chapter 11), although what became known as the Many Worlds Interpretation (MWI) of quantum mechanics is usually credited to the American physicist Hugh Everett (1930–82), who proposed it a few years later; he was unaware of Schrödinger's earlier work, and (unfortunately) did see the MWI in terms of the repeated "splitting" of reality into different branches. In my view, the most important thing to take away from the concept of the Multiverse, which is the only satisfactory way to explain why quantum computers work, is Schrödinger's idea that the wave function never collapses, and that all realities are (and always were) equally real. From there, it is but a short step to Schrödinger's other great interest.

Quantum physics and reality

The big problem with the Copenhagen Interpretation, even if you are comfortable with the idea of collapsing wave functions, has always been where you draw the line between the quantum world and the everyday world. In the classic thought experiment, it is implicitly assumed that it takes the consciousness of a human being to trigger the collapse and determine whether the cat is dead or alive. But isn't the cat itself competent to tell whether it is alive or dead? Could an ant trigger the collapse? Or a robot?

John Bell expressed this beautifully in an article in *Physics World* in 1990. He wrote: "What exactly qualifies some physical system to play the role of 'measurer'? Was the wave function of the world waiting to jump for thousands of years until a single-celled living creature appeared? Or did it have to wait a little longer, for some better qualified system [a person] with a Ph.D.?" What Bell set out as a *reductio ad absurdum* some scientists take literally. Steven Weinberg (b. 1933), interviewed in the November 2010 edition of *Scientific American*, said: "The universe may be like a giant Schrödinger's cat: When the cat is alive the cat knows he's alive (there are scientists to record what's going on in the universe); and in the other state, the cat doesn't know anything (there are no scientists to observe what's going on)." This is the Copenhagen Interpretation taken to its logical extreme—the Universe only exists because we are here to observe it. But I'm with Bell on this—what such a notion really indicates is the absurdity of the Copenhagen Interpretation.

There have been many attempts—apart from the Many Worlds Interpretation—to get round this problem by removing the role of the conscious observer. The most fashionable in recent years has been something called "decoherence." According to this idea, although a single quantum entity such as an atom can exist in a superposition of states, if this entity becomes entangled with a macroscopic object containing large numbers of atoms the complexity of the system forces the quantum superposition to "decohere" as the information describing the individual quantum entity gets lost among the vast number of possible combinations of interactions between all of the atoms in the macroscopic system. Put more simply, the decoherence occurs because there is no such thing as an isolated quantum entity; everything is entangled with the outside world. This has always seemed to me at best a variation on the Copenhagen Interpretation and at worst mumbo jumbo. If a single atom can exist in a quantum state but my desk, say, cannot, where do you draw the line? Will two atoms form a quantum system? Experiments show they do. So how many atoms does it take to make the system decohere? Three? Seventeen? Forty-two? It is exactly the same, but on a smaller scale, as the Schrödinger's Cat puzzle.

But the idea of decoherence does suggest another way of looking at the world, one more in keeping with Schrödinger's philosophy. Proponents of decoherence, looking from the outside in, say that entanglement makes the quantum world classical; looking from the inside out, however, it is equally valid to say that entanglement makes the classical world quantum. And in recent years experimenters have detected an increasing number of examples of quantum effects—in particular, entanglement—operating in macroscopic systems. It seems that there is no "outside world."

One of the pioneers of this line of work, Oxford-based physicist Vlatko Vedral (b. 1971), summarized the evidence in the June 2011 issue of *Scientific American*. Throughout the first decade of the twenty-first century, various researchers carried out a series of experiments in which they studied entanglement in increasingly large samples of material. In some of the earliest studies, crystals were tweaked by magnetic fields. In essence, the atoms in the samples were seen to line themselves up with the magnetic fields more quickly than they could have without the help of entanglement. The early versions of these experiments required very low temperatures, close to absolute zero (minus 273°C), to avoid interference by the jiggling of atoms due to thermal motion; but by the end of the decade various kinds of entanglement experiments were being carried out at temperatures well above 0°C.

The most impressive evidence of macroscopic entanglement so far comes from studies of migratory birds. In some species, the eye of the bird contains a type of molecule in which two electrons form an entangled pair. When this molecule is exposed to light, the entangled electrons become sensitive to magnetic fields. Experiments confirm that this changes the chemistry of the processes affecting vision. It has been suggested, but not yet proved conclusively, that the alteration in the chemistry of the photo-receptors means that an image of the magnetic field is produced in the bird's brain—that it can see the magnetic field.

Other evidence of entanglement at work in living systems comes from the chemistry of photosynthesis. Vedral says that it is no longer possible to accept that "large numbers of particles spontaneously behave classically . . . experiments now leave very little room for such processes to operate . . . few physicists now think that classical physics will ever really make a comeback on any scale." And "the implications of macroscopic objects such as us being in quantum limbo [are] mind-blowing."

And if there are no boundaries between the macroscopic world and the quantum world, if there is no collapse of the wave function, and if everything is entangled with everything else, the physicist's view of the world is, after all, not so different from Schrödinger's Vedantic vision of reality.

Postscript

Quantum Generations

In the summer of 2011, after completing the main text of this book, I was able to meet Terry Rudolph and discuss the foundations of quantum physics with him in his office, perched twelve floors up at Imperial College, with spectacular views across London. First, he told me about the circuitous route by which he had arrived there, and the surprise he experienced when he learned who his grandfather was.

Terry was born in Zimbabwe in 1973, and spent his childhood in southern Africa during a time of political turbulence. His immediate family (he has a brother and two sisters) moved to Malawi in 1979, but his grandmother stayed in Zimbabwe, and although he saw her once or twice, moving around southern Africa at that time was, as he puts it, "problematic"; after 1979, he said, "most of my memories of her are via letters." When the Rudolph family moved on to Australia in the mid-1980s, she stayed in Zimbabwe, and died about ten years later.

As a teenager in Australia, Terry was much more interested in sport (particularly squash) than academic studies, and spent many hours (up to six a day) training when he should have been studying, in the hope of becoming a professional. "I was very competitive," he says, "I guess I still am." In spite of his sporting obsession, he graduated from the University of Queensland in 1993, majoring in mathematics and physics, having chosen physics because it seemed like the hardest subject—though he points out that his father (now retired) was a chemist and teacher, and played a big part in his becoming a scientist. It was in the final weeks of his first "ordinary" degree that Terry became hooked on the fundamental puzzle of the foundations of quantum physics. It happened when one of the professors gave a lecture—the last lecture of his degree course—on Bell's inequality and the Aspect experiment. This was Terry's introduction to the idea of entanglement and non-locality, and (like many others) at first he could not believe it. He spent two weeks trying to find the flaw in the argument, so obsessed with it that he ignored his revision and nearly failed his final exam as a result. Having failed to find a flaw, he realized that here was something really difficult and profound, something worth making the effort to try to get to grips with; before this, he had really found physics too easy to be interesting.

The immediate result of this new obsession was a thesis leading to an honours degree from Queensland in 1994. Then, it seemed like a good time to take a year out and travel the world before settling down. Terry knew that his mother had a half-sister in Austria, Ruth Braunizer, and naturally he planned to visit her on the European leg of his travels. It was at this point that his mother realized it was time she told Terry who his grandfather was. "We were having breakfast, and my mum just said, 'There's something you should know. My father was Erwin Schrödinger.'" The news came as a complete surprise to the twenty-one-year-old would-be quantum physicist, who had just written his first scientific paper. As he put it to me, "I'd even read your book [*In Search of Schrödinger's Cat*] before I knew who my grandfather was."

After his gap year, which included meeting up with the Austrian branch of the family, Terry obtained a PhD from York University in Canada (awarded in 1998) and worked at the University of Toronto, the University of Vienna (collaborating with, among others, Anton Zeilinger), and Bell Laboratories, before settling in 2003 at Imperial, where he is now Co-Director, Centre for Doctoral Training in Controlled Quantum Dynamics. This means that he is in charge of some of the brightest students around, addressing the foundations of quantum theory, although he is also interested in practical applications, such as quantum computing. He is now easy about being Schrödinger's grandson, and neither makes any fuss about it nor finds it uncomfortable; but he makes a point of never reading any biographies of his grandfather (including this one) because he doesn't want to be influenced, consciously or unconsciously, by an awareness of his antecedents. "I don't want to second-guess myself," he says. Terry Rudolph is very much his own man, and although happy enough to bring me up to speed on his background, really wanted to talk about quantum reality—which was fine by me.

One aspect of this work which particularly intrigued me involves entanglement, thermodynamics, and the arrow of time. It is exactly the sort of thing that Schrödinger, with his deep appreciation of Boltzmann's work, might have been involved with if he were still around. Terry Rudolph and his colleague David Jennings have shown that it is hard to decide what is meant by an arrow of time for entangled "multipartite" systems, and have suggested ways to test these ideas using systems composed of a small number of qubits, where time might be seen to be going backwards, in the sense of decreasing entropy. Going beyond those simple experiments into an area of special interest to me with my background in cosmology, they have speculated that in the early, dense Universe the extreme conditions might have "allow[ed] any random physical interactions to exploit the correlations present, producing a gradual disappearance of the thermodynamic arrow the closer we get to the initial state of the Universe."[1] In other words, there was no time in the Big Bang!

But the most important work that Terry is involved with—and the most relevant to this book—concerns the possibility that Einstein and Bohm were right, and that there is an underlying "reality" which is described imperfectly by quantum mechanics. "I'm a fairly conservative physicist," he says. "In the community I work in, there are a lot of people who believe in things like Many Worlds, but I'm not prepared to entertain that until I am sure there's no other option." He also shares Schrödinger's disdain for the Copenhagen Interpretation, even if he wasn't influenced by his grandfather in reaching this position. Asked if he regards himself as interested in theory for theory's sake, or rather in practical applications like quantum computing, he replies: "Mainly, theory for theory's sake; but it's important to appreciate that quantum mechanics isn't just some abstract mathematical theory, but does have a bearing on reality. Quantum mechanics is about reality, it isn't about consciousness, or observers, or whatever. I think Einstein had a good point—quantum mechanics is incomplete. I'm sure there's something deeper."

The basic question is whether there is a one-to-one mapping between states of reality and quantum states—something Einstein discussed in correspondence with Schrödinger. In other words, is it possible for two (or more) quantum states to be associated with the same single state of reality—or, conversely, for two or more states of reality to be described by the same quantum state? "Maybe quantum mechanics doesn't capture everything that's really going on," says Terry:

> If I know the state of reality, is it possible for me to infer what the quantum state is? If it's not, then many different elements of reality can be associated with one quantum state. Moreover, the same element of reality can be associated with two quantum states. So you can have a distribution of reality elements corresponding to one quantum state, and a distribution of reality elements corresponding to another quantum state, and the two distributions may overlap, so that some of the elements of reality can be associated with both the quantum states. For those reality elements you can't be sure which quantum state applies. The overlap points are ambiguous in some sense—the description of reality is ambiguous.

For a long time, it proved impossible to find a mathematical description of quantum mechanics that allowed this kind of overlapping. In a classic demonstration of how science sometimes develops, Terry spent a long time trying to prove that such overlapping is impossible, and in doing so eventually found an example which worked:

> That example had one pathological feature. It has a fundamental kind of non-separability. Even if you have two systems that have never interacted, it isn't enough to know everything about system *A* and everything about system *B* separately in order to predict how they will behave in future. Theories in which the distributions overlap have a fundamental non-separability to them. It's entanglement, but not in the usual sense.

Terry is still trying to come to grips with the consequences of this discovery—at the moment it's just a mathematical proof. But it seems to rule out whole classes of hidden variables theories.

This is just one example of the way in which the foundations of quantum mechanics still—or perhaps I should say, once again—attract some of the brightest minds in science. There's a younger generation inspired by the ideas of quantum information theory. "I'm about the oldest of them," says Terry. "They're young guys, very enthusiastic, free from the old wrangling about interpretations and such like, and what they do has practical applications. A lot of progress is being made, but you don't always hear about it. It's an alive field."

What better note on which to end? Schrödinger always wanted a son, and saw the continuity of the genetic line as a kind of immortality; he would surely have been delighted to know about his grandson, and about his work at the frontiers of quantum mechanics.

Notes

Chapter 1: Nineteenth-Century Boy

1 Friedman, *An Unsolicited Gift.*

2 Unless otherwise indicated, comments about Schrödinger's early life come from unpublished manuscripts by Minnie and by Erwin himself, in the Schrödinger Archive at Alpbach; see also Moore, *Schrödinger: Life and Thought.*

3 Quoted by Mehra and Rechenberg, *The Historical Development of Quantum Theory.*

4 Quoted by Moore, *Schrödinger: Life and Thought.*

Chapter 2: Physics before Schrödinger

1 The term "scientist" had not been invented then, but it is the most apt.

2 In principle, you need to include electromagnetism as well, but the essence of the argument is still the same.

3 Rival "theories" like this are sometimes called "models"; I shall use the terms interchangeably in this book.

4 Although Boltzmann himself did cross the Atlantic at the end of the nineteenth century, to give a series of lectures, he did not meet Gibbs and remained unaware of the full significance of his work.

Chapter 3: Twentieth-Century Man

1 The need to prove this as late as 1905 is clear from Einstein's *Autobiographical Notes*, where he tells us that at that time he deliberately set out to find evidence "which would guarantee as much as possible the existence of atoms."

2 Information about Hansi Bauer and her relationship with Schrödinger comes from an interview she gave to Walter Moore quoted in *Schrödinger: Life and Thought*.

Chapter 4: The First Quantum Revolution

1 See Mehra and Rechenberg, *The Historical Development of Quantum Theory*, vol. 1.

Chapter 5: Solid Swiss Respectability

1 Strictly speaking, he only became a "von" in 1913, when a minor title was bestowed on his father.

2 From the version of his lecture published in 1929, as quoted by Mehra and Rechenberg, *The Historical Development of Quantum Theory*.

3 And not published in English until 1964.

4 He actually commented to Langevin that de Broglie had "lifted a corner of the great veil."

Chapter 6: Matrix Mechanics

1 One of the dreadful semi-puns that delight some scientists; Bohr = Beer.

2 This and the following quotations in this section are from Heisenberg's *Physics and Beyond*.

Chapter 7: Schrödinger and the Second Quantum Revolution

1 Published in *Physics Today*, 1976, vol. 29, no. 12.

2 Quoted in *Niels Bohr's Times*, where the author, Abraham Pais, comments: "To the best of my knowledge that was the last time Einstein wrote approvingly on quantum mechanics."

3 This was the work for which he would (eventually!) receive the Nobel Prize.

4 My emphasis.

5 Quoted by Pais, *Niels Bohr's Times* (Ehrenfest's emphasis).

6 Quoted by Pais, *Niels Bohr's Times*.

Chapter 8: The Big Time in Berlin

1 Dirac, *Directions in Physics*.

2 Hardly surprisingly, it didn't take much thought for London to decide to take up the offer, and he too became one of the Prof's protégés.

Chapter 9: The Coming of the Quantum Cat

1 See Born, *My Life*. A. J. Ayer's autobiography, *Part of my Life*, gives a deeper insight into the Oxford society of the time.

2 The paper, along with many other relevant papers, can be found in the volume edited by John Wheeler and Wojciech Zurek, *Quantum Theory and Measurement*; an extended discussion of the EPR Paradox can be found in the volume edited by Franco Selleri, *Quantum Mechanics versus Local Realism*. Einstein actually had very little to do with writing the paper, but allowed his name to go on it as co-author.

3 In another paper, published the same year in *Naturwissenschaften*, he coined the term "entanglement" to

describe the way the two quantum entities are linked. In his own words, "The characteristic of quantum mechanics, the one that enforces its entire departure from classical lines of thought [is that] by the interaction the two representatives (or ψ-functions) have become entangled."

Chapter 10: There, and Back Again

1 This chapter draws on *Schrödinger: Life and Thought* by Walter Moore, who had the opportunity to interview several of Schrödinger's contemporaries in these years of crisis who have since died, including Hermann Mark and Hansi Bauer-Bohm.
2 Quoted by Moore, *Schrödinger: Life and Thought.*
3 At the time we had this conversation, McCrea was preparing his contribution to the volume *Schrödinger*, edited by Clive Kilmister.
4 Quoted by Moore, *Schrödinger: Life and Thought.*
5 Many popular, and even academic, accounts imply that after Edwin Hubble discovered the redshift–distance relation at the end of the 1920s this was immediately accepted as an indication that the Universe is expanding. In fact, it took well over ten years for the idea of the Big Bang to catch on.

Chapter 11: "The Happiest Years of My Life"

1 There are two universities because originally Trinity College Dublin (TCD) was for Protestants, and the National University, of which University College is part, for Catholics; no Catholics attended TCD until the 1960s.
2 Quoted by Bitbol, *Schrödinger's Philosophy of Quantum Mechanics.* Other Schrödinger quotes in this section come from the same source unless otherwise indicated.

Chapter 12: What Is Life?

1 See Pauling's contribution to Kilmister, ed., *Schrödinger*.
2 It sounds even better in the original German—"Atomphysikalisches Modell der Mutation."
3 Cited by Max Perutz, in Kilmister, ed., *Schrödinger*.
4 Cited by Max Perutz, in Kilmister, ed., *Schrödinger*.
5 See my book *In Search of the Double Helix*.

Chapter 13: Back to Vienna

1 Letter quoted by Moore in *Schrödinger: Life and Thought*.
2 Quoted by Moore in *Schrödinger: Life and Thought*.
3 A military version of the honour, known as the Blue Max, existed prior to the end of the First World War.

Chapter 14: Schrödinger's Scientific Legacy

1 Interview in Davies and Brown, *The Ghost in the Atom*.
2 Interview in Bernstein, *Quantum Profiles*.
3 Bernstein, *Quantum Leaps*.
4 Bell, *Speakable and Unspeakable in Quantum Mechanics*.
5 Interview in *Omni* magazine, May 1988. Unknown to Bell at the time he found the silly mistake, the flaw in von Neumann's argument had also been pointed out by Grete Hermann, in 1935, but ignored.
6 Quoted by Aczel, *Entanglement*.
7 Quoted by Aczel, *Entanglement*.
8 Interview in Davies and Brown, *The Ghost in the Atom*.
9 Quoted by Gilder, *The Age of Entanglement* (you have to imagine Bell's soft Irish accent, especially on the word "unbelievable").
10 Quotes from *Science News*, 10 April 1993.
11 The details are in Zeilinger's book, *Dance of the Photons*.

12 With just 100 qubits, you would have the equivalent of 1,267 billion billion billion bits.

13 See Brown, *Minds, Machines, and the Multiverse*.

Postscript: Quantum Generations

1 *Physical Review* E, Vol. 81 (2010), p. 061130.

Sources and Further Reading

Archives

Archives for the History of Quantum Physics (AHQP), Science Museum Library, London

Einstein Archive, Princeton

Johns Hopkins University Archive

Oxford University Archive

Schrödinger Archive, Alpbach

Schrödinger Archive, Vienna

University of Berlin Archive

University of Wisconsin Archive

Vienna University Archive

Published sources

Aczel, Amir, *Entanglement* (Chichester: Wiley, 2003)

Al-Khalili, Jim, *Quantum: A Guide for the Perplexed* (London: Weidenfeld & Nicolson, 2003)

Ayer, A. J., *Part of My Life* (London: Collins, 1977)

Baggott, Jim, *Beyond Measure* (Oxford: Oxford University Press, 2004)

Baggott, Jim, *The Quantum Story* (Oxford: Oxford University Press, 2011)

Bell, John, *Speakable and Unspeakable in Quantum Mechanics* (Cambridge: Cambridge University Press, 1987)

Bernstein, Jeremy, *Quantum Profiles* (Princeton: Princeton University Press, 1991)

Bernstein, Jeremy, *Quantum Leaps* (Cambridge, Mass.: Belknap Press, 2009)

Bettelheim, Anton, et al., eds, *Neue Österreichische Biographie 1815–1918* (Vienna: Amalthea, 1957): includes contributions from Hans Thirring on Schrödinger and Hasenöhrl

Bitbol, Michel, *Schrödinger's Philosophy of Quantum Mechanics* (Dordrecht: Kluwer, 1996)

Blair, Linda, *The Happy Child* (London: Piatkus, 2009)

Bohr, Niels, *Atomic Theory and the Description of Nature* (Cambridge: Cambridge University Press, 1934)

Born, Max, *Natural Philosophy of Cause and Chance* (Oxford: Oxford University Press, 1949)

Born, Max, *The Born–Einstein Letters* (London: Macmillan, 1971)

Born, Max, *My Life* (London: Taylor & Francis, 1978)

de Broglie, Louis, *New Perspectives in Physics* (New York: Basic Books, 1962)

de Broglie, Louis, and Léon Brillouin, *Wave Mechanics* (London: Blackie, 1928)

Brown, Julian, *Minds, Machines, and the Multiverse* (New York: Simon & Schuster, 2000)

Campbell, Lewis, and William Garnett, *The Life of James Clerk Maxwell*, 2nd edn (London: Macmillan, 1884)

Cassidy, David, *Uncertainty: The Life and Science of Werner Heisenberg* (New York: Freeman, 1992)

Cercignani, Carlo, *Ludwig Boltzmann* (Oxford: Oxford University Press, 1998)

Cherfas, Jeremy, *Man Made Life* (Oxford: Blackwell, 1982)

Clare, George, *Last Waltz in Vienna* (London: Macmillan, 1980)
Cline, Barbara Lovett, *The Questioners* (New York: Crowell, 1965)
Cline, Barbara Lovett, *Men Who Made a New Physics* (Chicago: University of Chicago Press, 1987)
Crick, Francis, *Life Itself* (New York: Simon & Schuster, 1982)
Cropper, William, *Great Physicists* (Oxford: Oxford University Press, 2001)
Davies, Paul, and Julian Brown, *The Ghost in the Atom* (Cambridge: Cambridge University Press, 1986)
DeWitt, Bryce, and Neil Graham (eds), *The Many-Worlds Interpretation of Quantum Mechanics* (Princeton: Princeton University Press, 1973)
Dirac, Paul, *Directions in Physics* (New York: Wiley, 1978)
Einstein, Albert, *Autobiographical Notes*, ed. P. A. Schilpp (La Salle: Open Court, 1979)
Farmelo, Graham, *The Strangest Man* (London: Faber & Faber, 2010)
Feynman, Richard, *The Character of Physical Law* (London: BBC, 1965)
French, A. P., and P. J. Kennedy (eds), *Niels Bohr: A Centenary Volume* (Cambridge, Mass.: Harvard University Press, 1985)
Friedman, Dennis, *An Unsolicited Gift* (London: Arcadia, 2010)
Gamow, George, *Thirty Years that Shook Physics* (New York: Dover, 1966)
George, A. (ed.), *Louis de Broglie, Physicien et Penseur* (Paris: Albin Michel, 1953)
Gilbert, William, *De Magnete*, trans. P. Fleury Mottelay (New York: Dover, 1958; repr. of edn first publ. 1893)
Gilder, Louisa, *The Age of Entanglement* (New York: Knopf, 2008)

Gribbin, John, *In Search of Schrödinger's Cat* (London: Wildwood House, 1984; updated edn Black Swan, 2012)

Gribbin, John, *In Search of the Double Helix* (London: Penguin, 1995)

Gribbin, John, *Schrödinger's Kittens and the Search for Reality* (London: Weidenfeld & Nicolson, 1995; pb Phoenix, 1996)

Harrod, Roy, *The Prof* (London: Macmillan, 1959)

Heisenberg, Werner, *The Physical Principles of the Quantum Theory* (Chicago: University of Chicago Press, 1930)

Heisenberg, Werner, *Physics and Philosophy* (New York: Harper & Row, 1962)

Heisenberg, Werner, *Der Teil und das Ganze* (Munich: Piper, 1969)

Heisenberg, Werner, *Physics and Beyond* (London: Allen & Unwin, 1971)

Heisenberg, Werner, *Collected Works*, ed. W. Blum, H. P. Dür and H. Rechenberg (Berlin: Springer, 1984)

Hermann, Armin, *The Genesis of Quantum Theory* (Cambridge, Mass.: MIT Press, 1971)

Hermann, Armin, Karl von Meyenn and Victor Weisskopf (eds), *Wolfgang Pauli: Scientific Correspondence* (New York: Springer, 1979)

Hoffmann, Banesh, *The Strange Story of the Quantum* (London: Penguin, 1963)

Hutchins, Robert (ed.), *Gilbert, Galileo, Harvey* (Chicago: Encyclopedia Britannica, 1952): reprints in English of key works of each of these three pioneers of science

Jammer, Max, *The Conceptual Development of Quantum Mechanics* (New York: McGraw-Hill, 1966)

Jammer, Max, *The Philosophy of Quantum Mechanics* (London: Wiley, 1974)

Judson, Horace Freeland, *The Eighth Day of Creation* (London: Cape, 1979)

Kilmister, Clive (ed.), *Schrödinger* (Cambridge: Cambridge University Press, 1987)

Kragh, Helga, *Quantum Generations* (Princeton: Princeton University Press, 1999)

Lindley, David, *Boltzmann's Atom* (London and New York: Free Press, 2001)

Mahon, Basil, *The Man Who Changed Everything: The Life of James Clerk Maxwell* (Chichester: Wiley, 2003)

Mehra, Jagdish, and Helmut Rechenberg, *The Historical Development of Quantum Theory*, vol. 1 (in two parts): *The Quantum Theory of Planck, Einstein, Bohr, and Sommerfeld* (New York: Springer, 1982)

Mehra, Jagdish, and Helmut Rechenberg, *The Historical Development of Quantum Theory*, vol. 2: *The Discovery of Quantum Mechanics* (New York: Springer, 1982)

Mehra, Jagdish, and Helmut Rechenberg, *The Historical Development of Quantum Theory*, vol. 3: *The Formulation of Matrix Mechanics and Its Modifications 1925–1926* (New York: Springer, 1982)

Mehra, Jagdish, and Helmut Rechenberg, *The Historical Development of Quantum Theory*, vol. 4 (in two parts): *The Fundamental Equations of Quantum Mechanics 1925–1926* and *The Reception of the New Quantum Mechanics 1925–1926* (New York: Springer, 1982)

Mehra, Jagdish, and Helmut Rechenberg, *The Historical Development of Quantum Theory*, vol. 5 (in two parts): *Erwin Schrödinger and the Rise of Wave Mechanics* (New York: Springer, 1987)

Mehra, Jagdish, and Helmut Rechenberg, *The Historical Development of Quantum Theory*, vol. 6 (in two parts): *The Completion of Quantum Mechanics* (New York: Springer, part 1 2000, part 2 2001)

Moore, Walter, *Schrödinger: Life and Thought* (Cambridge: Cambridge University Press, 1989)

Olby, Robert, *The Path to the Double Helix* (London: Macmillan, 1974)

Pagels, Heinz, *The Cosmic Code* (London: Michael Joseph, 1983)

Pais, Abraham, *Subtle Is the Lord* (Oxford: Oxford University Press, 1982)

Pais, Abraham, *Inward Bound* (Oxford: Oxford University Press, 1986)

Pais, Abraham, *Niels Bohr's Times* (Oxford: Clarendon Press, 1991)

Pauli, Wolfgang (ed.), *Niels Bohr and the Development of Physics* (London: Pergamon, 1955)

Pauling, Linus, and Peter Pauling, *Chemistry* (San Francisco: Freeman, 1975)

Planck, Max, *Scientific Autobiography and Other Papers*, trans. Frank Gaynor (London: Williams & Norgate, 1950)

Price, W. C., S. S. Chissik and T. Ravensdale (eds), *Wave Mechanics: The First Fifty Years* (London: Butterworth, 1973)

Rae, Alastair, *Quantum Physics: Illusion or Reality?* (Cambridge: Cambridge University Press, 1986)

Santesson, Carl Gustaf (ed.), *Les Prix Nobel en 1933* (Stockholm: Norstedt, 1934); see also http://nobelprize.org/nobel_prizes/physics/laureates/1933/schrodinger-speech.html

Schrödinger, Erwin, *Four Lectures on Wave Mechanics* (London: Blackie, 1928)

Schrödinger, Erwin, *Collected Papers on Wave Mechanics* (London: Blackie, 1928)

Schrödinger, Erwin, *Science and the Human Temperament* (London: Allen & Unwin, 1935)

Schrödinger, Erwin, *What Is Life?* (Cambridge: Cambridge University Press, 1944)

Schrödinger, Erwin, *Statistical Thermodynamics* (Cambridge: Cambridge University Press, 1946)

Schrödinger, Erwin, *Space–Time Structure* (Cambridge: Cambridge University Press, 1950)

Schrödinger, Erwin, *Science and Humanism: Physics in our Time* (Cambridge: Cambridge University Press, 1951)

Schrödinger, Erwin, *Nature and the Greeks* (Cambridge: Cambridge University Press, 1954)

Schrödinger, Erwin, *Expanding Universes* (Cambridge: Cambridge University Press, 1956)

Schrödinger, Erwin, *Science, Theory and Man* (New York: Dover, 1957)

Schrödinger, Erwin, *Mind and Matter* (Cambridge: Cambridge University Press, 1959)

Schrödinger, Erwin, *Biographical Memoirs of the Royal Society* (London: Royal Society, 1961)

Schrödinger, Erwin, *My View of the World* (Cambridge: Cambridge University Press, 1964)

Schrödinger, Erwin, *Letters on Wave Mechanics* (New York: Philosophical Library, 1967)

Schrödinger, Erwin, *The Interpretation of Quantum Mechanics* (Dublin seminars 1949–55 and other unpublished texts), ed. Michel Bitbol (Woodbridge, Conn.: Ox Bow Press, 1995)

Schweber, Silvan, *QED and the Men Who Made It* (Princeton: Princeton University Press, 1994)

Scott, William, *Erwin Schrödinger: An Introduction to His Writings* (Amherst: University of Massachusetts Press, 1967)

Selleri, Franco (ed.), *Quantum Mechanics versus Local Realism: The Einstein–Podolsky–Rosen Paradox* (New York: Plenum, 1988)

Shearer, J. F. (trans.), *Collected Papers on Wave Mechanics* (London and Glasgow: Blackie, 1928) van der Waerden, B. L. (ed.), *Sources of Quantum Mechanics* (New York: Dover, 1968)

Weber, Robert, *Pioneers of Science*, 2nd edn (Bristol: Adam Hilger/Institute of Physics, 1988)

Westfall, Richard, *Never at Rest: A Biography of Isaac Newton* (Cambridge: Cambridge University Press, 1980); a shorter version of this book was published by Cambridge University Press in 1993 under the title *The Life of Isaac Newton*

Wheeler, John, and Wojciech Zurek (eds), *Quantum Theory and Measurement* (Princeton: Princeton University Press, 1983)

Woolf, Harry (ed.), *Some Strangeness in the Proportion* (Reading, Mass.: Addison-Wesley, 1980)

Zeilinger, Anton, *Dance of the Photons* (New York: Farrar, Strauss & Giroux, 2010)

Index

action at a distance, 154, 165, 178, 220, 268, 274, 277
Aharonov, Yakir, 264
Aigentler, Henriette von, 39
Alpbach, 218, 246, 248, 254, 256–8
American Philosophical Society, 181
Anderson, Carl, 159
Annalen der Physik, 55, 129, 132, 133
Annales de physique, 110
anti-particles, 159–60
Arosa, 97, 101, 125, 126
Arzberger, Hans, 10
Arzberger, Rhoda (née Bauer, aunt), 9, 10–11
Aspect, Alain, 266–9
atoms: Bohr model, 83–6, 87, 97, 112; Boltzmann's work, 38, 39, 40; concept, 38, 42, 55, 78; Copenhagen Interpretation, 154; decoherence, 282; Einstein's work, 55, 75, 87–9, 109; entanglement experiments, 283; "green pamphlet," 230, 233; Mach's view, 42; Maxwell's work, 37; nuclear model, 81–2; Planck's work, 67, 73–4, 89; Poincaré's work, 40; quantum chemistry, 228–9; quantum computing, 278, 280; quantum physics, 2, 78; quantum spin of electron, 114, 125; quantum teleportation experiments, 277; Rutherford's work, 79, 80, 81–2, 84, 86; Schrödinger's work, 49, 51, 64, 99–101, 125, 130, 135, 234; structure, 84
Austria: *Anschluss* (1938), 191–2, 193; army, 47, 51, 54; First World War and aftermath, 4, 54–7, 59, 60–1, 104, 243; International Atomic Energy Agency representation, 5; Nazism, 188, 192–4; religion, 10, 15, 192; Schrödinger's flight from, 195–6; Schrödinger's return to, 183–4, 185, 187, 188, 219, 243–4, 247; Second World War aftermath, 217, 243–5
Austria-Hungary, 14, 53, 59, 60
Austrian Empire, 13–14, 15, 169, 187, 190
Austrian Physical Society, 253

Baird, John Logie, 274
Ballot, Christoph Buys, 15–16
Bamberger, Emily (Minnie, née Bauer, aunt), 9, 10, 11, 12, 18
Bamberger, Helga (cousin), 10
Bamberger, Max, 10
Bär, Richard, 195, 197
Bauer, Alexander (grandfather), 8, 9–10, 12, 64
Bauer, Alexander (great-grandfather), 8

Bauer, Emily (Minnie, aunt), *see* Bamberger
Bauer, Emily (Minnie, née Russell, grandmother), 8–9, 10
Bauer, Friedrich (Fritz), 61–2
Bauer, Georgie, *see* Schrödinger
Bauer, Johanna (Hansi), *see* Bohm
Bauer, Josepha (née Wittmann-Denglass, great-grandmother), 8
Bauer, Rhoda (aunt), *see* Arzberger
BBC, 176–7, 217
Becquerel, Henri, 78–9
Bell, John, 262–3, 264–6, 273, 278, 281–2
Bell's inequality, 264–5, 268
Bennett, Charles, 271, 272, 273, 274, 275, 277
Berlin: Academy of Sciences, 73; Kaiser Wilhelm Institute for Chemistry, 224–5, 231, 232; Schrödinger's departure, 168–9, 192; Schrödinger's professorship, 144, 147, 148–9, 150–2, 154, 156–7, 176, 256; Schrödinger's work, 160; University of, 38, 70, 95, 139, 150–2, 154–6
Bernstein, Jeremy, 262
Bertel, Annemarie (Anny), *see* Schrödinger
Besso, Michele, 130
birds, vision, 283
Bitbol, Michel, 219–20
Blackett, Patrick, 139
Blair, Linda, 11
Bloch, Felix, 124, 129
Bohm, David, 260–1, 262–3, 288
Bohm, Franz, 170, 183
Bohm, Johanna (Hansi, née Bauer): escape from Austria to London, 198; escape from Germany to London, 183; marriage, 170; memories of Schrödinger, 62; pregnancy, 185; relationship with Schrödinger, 170, 183, 185, 193–4, 198, 214, 215, 216; in Vienna, 185, 191
Bohr, Niels: on collapse of wave function, 145, 165, 182; on complementarity, 143, 144, 152; Copenhagen Institute, 122, 139, 154, 231; Copenhagen Interpretation, 122, 144–7, 154, 260; Einstein's views of his work, 153, 178; Festival, 112; honours, 78, 189; influence, 4, 78, 82, 260; "Light and Life" lecture, 231–2; model of the atom, 83–6, 87, 97; Nobel Prize, 78; quantization rules, 124; relationship with Schrödinger, 136–7; Schrödinger's views of his work, 137, 144–5, 165, 218, 255; work with Heisenberg, 112–13, 138–9, 142, 143–4, 152
Boltzmann, Ludwig: background, 38; career, 38–9, 41–3, 44–5; depression, 39, 42; education, 16, 18, 38; on entropy, 237; influence on Schrödinger, 44, 51, 99, 288; on international nature of physics, 17; marriage, 39; relationship with Mach, 41–2; research, 21; statistical approach, 39, 40–1, 44, 70, 72, 160; Stefan–Boltzmann Law of black body radiation, 16; suicide, 43; work on atoms, 42, 55; work on thermodynamics, 22, 39, 41, 70, 72, 99, 160
Born, Max: background and education, 118; in Cambridge, 173; career, 112, 149, 150, 154, 167, 185–6; on chance and probability, 122; on Copenhagen school, 122–3, 142; on Dirac's work, 119–20; Edinburgh professorship, 185–6; in Göttingen, 112; Heisenberg's studies, 112, 114, 117–18, 121–2; in Italy, 169; matrix mechanics, 117–19; *Natural Philosophy of Cause and Chance*, 261–2; Nobel controversy, 121–2; Nobel Prize,

122; on quantum mechanics, 102, 138; quantum revolution, 120; relationship with Schrödinger, vii, 175, 195, 198, 206, 256–7; retirement, 186; sacked under Nazis, 167; Schrödinger's response to his work, 132–3, 135; statistics, 144; on von Neumann's work, 261–2; work on wave function, 135, 228
Bose, Satyendra Nath, 102, 108–10
Bose–Einstein statistics, 109, 114
bosons, 109
Bragg, Lawrence, 49
Bragg, William, 49
Braunizer, Andreas (grandson), 252
Braunizer, Arnulf, 247, 252
Braunizer, Ruth (née March, daughter): in Belgium, 200; birth, 175; birth of son, 252; care of Arthur, 247, 251; in Dublin, 204, 210, 246; in Graz, 189; half-sisters, 211–13, 286; in Innsbruck, 183, 184, 185, 213, 217, 246, 247; marriage, 247; in Oxford, 175; pregnancy, 247, 251; relationship with Anny, 189, 216; relationship with father, 217, 251; relationship with mother, 189, 247, 251
Brecht, Bertolt, 155
Breslau, 65, 66, 118
Bristol University, 231
Brown, Robert, 55
Browne, Monsignor Paddy, 205, 206, 207, 237
Brownian motion, 55
Bunsen, Robert, 82

Cahill & Company, 237–8
California: Institute of Technology, 150; University of, 43
Cambridge: Born in, 173; Cavendish Laboratory, 36–7, 79, 240; Heisenberg's lectures, 117, 119; Maxwell at, 34, 36; Newton at, 25–6; Philosophical Society, 179; Schrödinger's visits, 214; Tarner Lectures, 248
Cambridge University Press (CUP), 215, 238, 248
"Can Quantum Mechanical Description of Physical Reality Be Considered Complete?"(EPR paper), 177
CERN, 262, 263
Charles, Emperor, 53–4, 57, 58, 60
Chemical and Engineering News, 240
Chemical–Physical Society, 253
Chicago, 149, 150
chromosomes, 211, 227, 233, 235–6, 241
Clauser, John, 266
Clausius, Rudolf, 92, 93
cloud chamber, 138–9, 140
Cockcroft, John, 210
Cold War, 243
colour vision, 4, 63–4, 101
Columbia University, 150
Como conference (1927), 144, 146
complementarity, 143, 144, 234, 255
Compton, Arthur, 89–90, 102
Compton effect, 90
Condon, Edward, 229
Congress of Vienna (1913), 52–3
Copenhagen, Niels Bohr Institute, 122, 136–8, 139, 144, 154
Copenhagen Interpretation, 143–7; author's view, 257, 282; Bell's view, 262, 281–2; Bohm's view, 260; Born on, 122–3, 142; consensus view, 142, 153, 154; Einstein's view, 144–5, 155, 177–8, 180, 254, 260; origins, 122, 142; package, 144–5, 154; predictions, 165; presentation, 146–7; probability in, 161, 162; Schrödinger's view, 147, 153, 155, 220, 221, 247, 254–5, 257, 259, 288
Cosmic Code, The (Pagels), 134
cosmic rays, 51, 55, 159, 159
cosmology, 189, 199, 288
Cramer, John, 162, 163, 164

Crick, Francis, 4, 239–41, 242
cryptography, 269, 270–4
Curie, Marie and Pierre, 79, 80
Czernowitz, 60

D-Wave Systems, 279
Darwin, Charles, 19
Darwin, Charles (grandson of above), 184–5
Davies, Paul, 265
Davy, Humphry, 33
de Broglie, Louis: background and education, 127; Einstein's view of his work, 110, 129; influence, 4, 110–11, 124–5, 128, 132, 158; pilot wave model, 152, 260–1; Solvay Congress, 146, 152; thesis, 110–11, 124, 127–8
de Broglie, Maurice, 127
De magnete (Gilbert), 23, 25
De motu corporum in gyrum (Newton), 26–7
de Valera, Éamon ("Dev"): background and career, 201–3; Dublin Institute for Advanced Studies, 5, 195, 198, 203–4, 207; invitation to Schrödinger, 5, 195, 197–8, 200, 203–4; Schrödinger's departure, 247; Schrödinger's lectures, 211
de Valera, Sinéad (née Flanagan), 202
Debye, Peter, 94, 95, 102, 124–5, 129, 189
decoherence, 282
Delbrück, Max, 224–5, 227, 230–4, 238, 239
Descartes, René, 25
Deutsch, David, 261, 280
Dewar, Katherine Mary, 34
Dieks, Dennis, 270
diffusion equation, 160, 161
Dirac, Paul: appearance and character, 120, 156, 210; in Cambridge, 214; Dirac Equation, 158–60; *Directions in Physics*, 120–1, 154; Dublin colloquium, 210; education, 119–20; Fermi–Dirac statistics, 109, 114; influence, 4, 120; Nobel Prize, 121, 171; Solvay Congress, 146, 156; transformation theory, 133–4; view of interpretation, 154; work on quantum mechanics, 119–20
Dirac Equation, 158–60
Directions in Physics (Dirac), 120–1, 154
DNA, 4, 225–7, 234, 239, 241
Dollfuss, Engelbert, 188
Doppler, Johann Christian, 15–16
Doppler effect, 16
Dora (cousin), 11
double slit experiment, 31, 63, 145–6, 153, 164
Dublin: Austrian community, 208; Institute for Advanced Studies (DIAS), 5, 195, 198, 201, 203, 206–7, 209, 214, 215, 254; Schrödinger in, 201, 204–11, 218–19, 244; Schrödinger "family life" in, 211–14, 215–16; Schrödinger's arrival, 198, 200; Schrödinger's departure, 219, 246–7; Trinity College (TCD), 206, 207, 209–10, 211; University College, 205, 207, 211

Eckart, Carl, 133
Eddington, Arthur, 189–90, 210
Edinburgh University, 184–5
Ehrenfest, Paul, 147
Einstein, Albert: on "action at a distance," 153–4, 220; *annus mirabilis*, 130; Berlin professorship, 148, 155, 156; Bose–Einstein statistics, 109, 114; Bose's work, 108–9; childhood, 12; Congress of Vienna, 52–3; Cramer's work, 163–4; on de Broglie's work, 110, 127, 129; on double slit experiment, 153–4; education and career, 75, 93–5, 148; EPR

Paradox, 177–9, 265; experiences of anti-Semitism, 105, 158; experiences of Nazism, 166–7; on Feynman's work, 163; general theory of relativity, 58–9, 85, 86, 95, 189; influence on Heisenberg's work, 139–40; influence on Schrödinger's work, 58–9, 99, 101, 102, 109–10, 132, 181; on Mount Wilson experiment, 105; Nobel Prize, 67, 78; at Princeton, 5, 177, 203; relationship with Schrödinger, 155, 156, 180, 184; on Schrödinger, 130; on Schrödinger's cat, 182; Solvay Congress, 146, 152, 153; special theory of relativity, 3, 22, 44, 75, 93, 101, 104; on underlying "reality," 178–9, 288–9; view of chance and probability, 101, 122; view of Copenhagen Interpretation, 144–5, 155, 177–8, 180, 254, 260; work on Brownian motion, 55; work on light quanta, 3–4, 63, 67, 75–8, 83, 102, 127; work on quantum theory of radiation, 86–90
Ekert, Artur, 272–3
electromagnetic oscillators, 72, 73
electromagnetism, 16, 22, 33, 34–7, 210
electron(s): Bohr's work, 84–5, 86, 97, 143, 145–6, 153; bond, 229–30; Born's work, 135; "collapse of the wave function," 145; Compton's work, 89; Copenhagen Interpretation, 145–6, 153, 161; Copenhagen scientists' work, 138; de Broglie's work, 114, 125, 128, 152; Dirac's work, 158–9, 160; Einstein's work, 87, 88; energy, 76, 77; entangled, 283; Fermi–Dirac statistics, 109, 114; Feynman's work, 162–3; Heisenberg's work, 114; interaction between, 163; Lenard's work, 76; measurement of charge on, 55, 77; Millikan's work, 55, 77; negative, 159, 160; orbits, 64, 84–5, 97, 114, 118; pilot wave model, 152, 260–1; quantum teleportation, 276; quantum "transaction," 163–5; radiation resistance, 162–3; Rutherford's work, 79–82; Schrödinger's work, 97–8, 125, 128, 129, 135–7, 152–3, 160–1, 220; sharing of, 228; Solvay Congress (1927), 146, 152–3; spin, 114, 125, 158–9, 177–9, 264; "spooky action at a distance," 178; Thomson's work, 79; trajectory in cloud chamber, 138–41, 220; wave equation, 125, 128, 129, 135–6, 261; waves and particles, 110, 125, 128, 135–6, 143, 145, 228, 261
entanglement: decoherence, 282; experimental confirmation of, 259, 267–8; FTL signalling, 269–70; macroscopic, 283–4; quantum computing, 278–9; quantum cryptography, 270, 273; quantum teleportation, 274, 276–7; Rudolph's work, 286, 287–8, 290; Schrödinger's work, 181, 221; term, 181, 259
entropy, 93, 109, 236–7, 253, 288
EPR Paradox, 177–9, 259, 263, 265, 272–3, 275–6
Epstein, Paul, 65, 95, 96
ETH (Eidgenössiche Technische Hochschule), *see* Zürich
Ettinghausen, Andreas, 16
Everett, Hugh, 280
evolution, 19, 224, 227, 249
Exner, Franz, 17, 48–9, 52, 99

Faraday, Michael, 33, 34, 36
Farmelo, Graham, 120
faster-than-light (FTL) communication (signalling), 153–4, 178–9, 268–70

Fermi, Enrico, 196
Fermi–Dirac statistics, 109, 114
fermions, 109
Feynman, Richard, 31, 162
First World War, 53–9, 127
Forster family, 8–9
Fowler, Ralph, 119
Franklin, Rosalind, 241
Franz Ferdinand, Archduke, 53–4
Franz Josef, Emperor, 13–14, 53, 57
Fraunhofer, Josef von, 82
free will, 2, 3, 29–30, 237
Fresnel, Augustin, 31, 32–3
Friedman, Dennis, 11
Friedrich Wilhelm III, King of Prussia, 155
Frimmel, Franz, 45

Galileo Galilei, 24–5, 27, 42
gas in sealed box, 40–1
gases, kinetic theory, 37, 38, 39
Geiger, Hans, 81
genes: changes in, 211, 227, 232; copying process, 226, 234, 241; Delbrück's work, 230, 232–3, 234; DNA, 225–7, 234, 241; Haldane's suggestion, 234; molecules, 230, 232–3; Schrödinger's work, 211, 230, 234, 235–6, 240
Ghent, University of, 198–200
Gibbs, Willard, 39
Gilbert, William, 23–4, 25
Gordon, George, 192, 194
Göttingen, 112–13, 114, 118–19, 154
gravity: Einstein's work, 53, 58, 85, 95; Newton's work, 1–2, 23, 26, 27–9, 53, 85; Schrödinger's work, 210
Graz: Boltzmann at, 38–9; Nazism, 187–8, 190, 191, 192, 194; Schrödinger's dismissal, 195, 196; Schrödinger's lectures, 190; Schrödinger's professorship, 183–4, 185–6, 187, 188, 194; Schrödinger's research, 189
"green pamphlet," 227, 230–4
Greene, Blathnaid Nicolette (Schrödinger's daughter), 212
Greene, David, 208, 211–12

Habicht, Conrad, 75
Habsburg family, 13
Haldane, J. B. S., 234
Halifax, Lord, 191, 192
Halley, Edmond, 26
Hamilton, William, 119
Hasenöhrl, Friedrich (Fritz), 16, 17, 44–6, 48, 49, 57
Heidelberg, 38, 105, 118
Heisenberg, Werner: concept of half-integer quantum numbers, 113; concept of quantum uncertainty, 140–2, 143–4, 147; Copenhagen Interpretation, 144, 147; education and career, 112–13, 144, 148, 154, 173, 174; influence, 4; matrix mechanics, 111, 112, 117–18, 119, 132–3, 134, 139; Nobel Prize, 121–2; *Physics and Beyond*, 112–13; relationship with Schrödinger, 136–7, 138; Schrödinger's view of his work, 132–3, 142; Solvay Congress, 146–7, 152–3; theory of quantum world, 35, 111; work on quantum jumps, 114–17
Heisenberg's Uncertainty Principle, 140–2, 144
Heitler, Walter, 207–8, 214, 229
Heligoland, 114–15
Hemingway, Ernest, 56
Herbert, Nick, 269
heredity, 107, 224
Hertz, Heinrich, 37
Hess, Victor, 51–2
Hibben, John, 174
hidden variables, 261, 262, 263, 264, 290
Hindenberg, Field Marshal, 155, 166
Hiroshima, nuclear bombing, 214, 238
Hitler, Adolf: Austrian policy, 187–8, 190, 191; defeat of France, 206;

imprisonment, 104; invasion of Austria, 191; invasion of Soviet Union, 208; letter to Schrödinger, 176; rise to power, 121, 166
Hooke, Robert, 25, 26, 30–1
Hoover, Herbert, 61
Humboldt, Alexander von, 155
Huygens, Christiaan, 30–1
hydrogen atom, 125, 229

IBM Research Center, 271
ICI (Imperial Chemical Industries), 167, 168, 170, 175, 183
imaginary numbers, 98
In Search of the Multiverse (Gribbin), xi, 278
In Search of Schrödinger's Cat (Gribbin), 85, 111, 196, 269, 287
Innitzer, Cardinal, 192
Innsbruck: March household, 157, 183, 189, 200, 213, 216, 246, 247, 251; meeting of German scientists (1924), 103–4, 109; professorship offer, 104; quantum teleportation experiments, 277; Schrödinger at, 217–18
interference pattern, 32
International Atomic Energy Agency, 5
Inward Bound (Pais), 126
Irish Times, 209
Italian Physical Society, 245
Ithi, *see* Junger

Jeans, James, 71, 74
Jena, 62, 64
Jennings, David, 288
Jews: anti-Semitism, 105; in Austria, 15; Einstein's career, 105, 166; Hansi's background, 183, 194; Lindemann's aid to scientists, 167–8; Nazism in Austria, 191–2, 244; Nazism in Germany, 166–7; Schrödinger's position, 105, 166, 193–4
Johns Hopkins University, 150
Jordan, Pascual, 118–19, 120–2, 133, 134
Jung, Carl, 215, 250
Junger, Itha (Ithi), 131–2, 157–8, 168, 177

Kavanagh, Patrick, 208
Kepler, Johannes, 26
Khrushchev, Nikita, 244, 245
kinetic theory, 37, 38, 39
King's College, London, 241
Kirchhoff, Robert, 68, 82–3
Klein, Oskar, 144
Klimt, Gustav, 21
Kohlrausch, Fritz, 45, 52
Kolbe, Ella, 46
Krauss, Felicie, 50–1, 52
Krauss, Karl and Johanna, 50

Landé, Alfred, 113
Langevin, Paul, 110
Laplace, Pierre-Simon, 235
Large Hadron Collider, 220
lasers, 89
Laue, Max von: career, 65, 94, 95, 96, 156; Schrödinger's visit, 184; view of Copenhagen Interpretation, 254; work on X-ray crystallography, 49, 53, 94
laws of motion, 27–9
Lean, Lena, 210, 213
Lehrbuch der Physik, 64
Leiden, 65, 96, 118
Leipzig, 42, 144, 154
Lemaître, Georges, 199
Lenard, Philipp, 75–6, 77, 105
light: Einstein's work, 75–7, 87, 89, 99; faster-than-light communication, 153–4, 178–9, 268–70; momentum, 99–100, 102; Newton's work, 25, 30; particle theory, 30, 31, 32, 35, 63; Planck's work, 72; polarized, 271; quanta, *see* photons; Schrödinger's work, 99–100, 102; spectroscopy, 82–3; speed, 35, 89, 104, 153, 178, 267–8, 275, 277;

light: Einstein's work (*cont.*)
wave theory, 30, 31–2, 35, 63, 71, 82, 102, 128
Lindemann, Frederick Alexander, 167–8, 170, 173, 175, 197, 200
Listener, The, 176
Lockheed Martin, 279
Lockyer, Joseph, 83
London, Fritz, 168, 229
London: Imperial College, 285, 287; King's College, 241; University College, 215, 234, 239
Loschmidt, Josef, 16–17, 38

MacEntee, Barbara, 206
MacEntee, Máire, 206
MacEntee, Margaret (née Browne), 206
MacEntee, Seámus, 206
Mach, Ernst, 41–3
McCrea, William, 196, 200, 204, 207, 210, 217
Madison, 149–50
Madrid, 175, 176
magnetism, 23, 49
Many Worlds Interpretation (MWI), 222, 280, 282, 288
March, Arthur: death, 252; illness, 247, 251, 252; in Innsbruck, 189, 200, 213, 217; Italian holiday, 169; marriage, 157; Oxford post, 168, 172, 175; Princeton question, 174; relationship with Schrödinger, 170, 217; return to Innsbruck, 183
March, Hilde: Arthur's death, 252; Arthur's illness, 247, 251, 252; in Belgium, 200; birth of daughter, 175; birth of grandson, 252; in Dublin, 200, 204, 213; education, 208; in Graz, 189; marriage, 157; in Oxford, 172; pregnancy, 170, 174; relationship with Schrödinger, 168, 169–70, 174, 184, 189, 190–1, 195, 200, 204; return to Innsbruck, 183, 213
March, Ruth George Erica (Schrödinger's daughter), *see* Braunizer
Mark, Hermann, 190
Marsden, Ernest, 81
matrices, 117–19, 122, 132, 134
matrix mechanics, 111, 112, 118–19, 133–5
Maxwell, James Clerk: achievements, 37–8; background, 33–4; death, 36; education and career, 34–5, 36; influence, 38; marriage, 34; Maxwell distribution, 37; statistical techniques, 37–8, 67, 70; theory of electromagnetism, 22, 33, 35–7, 69, 84, 162, 210; *Treatise on Electricity and Magnetism*, 36; work on light, 35, 63, 128
May, Sheila, 208, 211–13
Meitner, Lise, 156, 231, 232, 253, 254
Mendel, Gregor, 224
Michelson, Albert, 104–5
Millikan, Robert Andrews, 55, 77–8, 189
Minkowski, Hermann, 93–4
molecules: arrangements of atoms, 229; Bohr's work, 85; Boltzmann's work, 99; Copenhagen Interpretation, 154; DNA, 225–6, 241; formation, 228; genes, 230, 232–3, 241; Heisenberg's work, 142; helical structure, 240–1; "of life," 226–7; Loschmidt's work, 16; macroscopic entanglement, 283; Maxwell's work, 37; quantum computing, 278; quantum teleportation, 275; RNA, 226; Schrödinger's work, 49, 51, 55, 58, 100, 130, 160–1, 230
momentum: Bohr's work, 143; Compton's work, 89–90; de Broglie's work, 128; Einstein's work, 89–90, 99; EPR Paradox, 177, 264; Heisenberg's work, 140–1; law of conservation of,

99–100; matrix mechanics, 119, 133; Newton's laws, 29; Schrödinger's work, 99–101, 102, 133
Moore, Walter, 62, 131, 170, 190, 212
Morgan, Thomas Hunt, 233
Morley, Edward, 105
Moscow Declaration (1943), 244–5
Mount Wilson experiment (1921), 105
Multiverse, 280–1
Mussolini, Benito, 188
mutation, 227, 232
Myles, *see* O'Nolan

Nagasaki, nuclear bombing, 214, 238
Napoleon, 31
National Academy of Sciences, US, 238
National University of Ireland, 206
Natural Philosophy of Cause and Chance (Born), 261–2
Nature, 41, 199, 231, 241, 270, 277
Nature of the Chemical Bond, The (Pauling), 229
Naturwissenschaften, Die, 181
Nernst, Walther, 156
New York, 149–50
Newton, Isaac: education and career, 25–6, 30, 34; laws of motion, 26–9, 40; laws of physics, 1–2, 22–3, 29; *Opticks*, 30; *Principia*, 27, 36; theory of gravity, 1, 26, 28, 29, 53, 85; work on light, 25, 30–1
Nobel Committee, 120, 121
Nobel Prize: Blackett (1948), 139; Bohr (1922), 78; Born (1954), 122; Cockcroft (1951), 210; Compton (1927), 89–90; Crick (1962), 241; Delbrück (1969), 231; Dirac (1933), 121, 171; Einstein (1921), 67, 78; Heisenberg (1932), 121, 122; Hess (1936), 51; Laue (1914), 94; Millikan (1923), 78; Pauling (1954), 229; Planck (1918), 78; Rutherford (1908), 80; Schrödinger (1933), 3, 7, 97, 171, 172; Walton (1951), 209–10; Watson (1962), 241; Wilson (1927), 138
Nolan, Kate, 213
nuclear physics, 156, 214

O'Brien, Conor Cruise, 206
O'Nolan, Brian ("Myles"), 208–9
Opticks (Newton), 30
Ortega, José, 176
Ostwald, Wilhelm, 42
Oxford, 4–5, 170–1, 175, 195, 198

Pagels, Heinz, 134
Pais, Abraham, 126
particle mechanics, 151
particle theory, 35, 214
particles: alpha, 79, 80, 81; anti-particles, 159–60; beta, 79; Bohr's work, 143; Born's work, 136; Bose's work, 108; bosons, 109; Copenhagen Interpretation, 145, 153, 247, 255; Crick's work, 240; de Broglie's work, 110, 124, 128, 152; Einstein's work, 63, 76, 77, 89, 99; entangled, 276–7, 284; fermions, 109; Heisenberg's work, 111; Maxwell's work, 84; momentum, 89, 90, 99; negatively charged (electrons), 79, 86; Newtonian physics, 2, 29–30; Newton's work on light, 30, 31; number in Universe, 190; phase space, 153; photons, 109; positively charged (protons), 79, 86; quantum chemistry, 228; quantum teleportation, 276–7; quantum transaction, 163; radiation resistance, 162–3; Schrödinger's work, 63, 99, 136, 220–1; Solvay Congress, 216; spin, 114, 125; "spooky action at a distance," 178; statistical mechanics, 39; subatomic, 100,

particles (*cont.*)
101; trajectory in cloud chamber, 138–9, 220; waves and, 3, 32, 35, 63, 77, 136, 143, 247, 255, 261; Young's work, 32
Pauli, Wolfgang: career, 154, 231; Dublin visit, 214; Feynman's Princeton talk, 163; on half-integer quantum numbers, 113; Heisenberg's letters, 118, 141; on matrix mechanics and wave mechanics, 133; on measurement of atom, 115; Solvay Congress, 146
Pauling, Linus, 229–30, 233–4, 240
phase space, 153
Philosophical Magazine, 108
photons: Aspect's experiments, 267–9; Bell's work, 264, 266, 267; Bose–Einstein statistics, 109, 114; Bose's work, 108; Clauser's experiment, 266; clones of, 269–70, 274; Compton's work, 89–90, 102; Einstein's work, 67, 75, 76–7, 87–9, 90, 102; entangled, 267–8, 273, 276, 277; green pamphlet on, 230; light quanta, 67, 75, 76; momentum, 102; Planck's work, 67, 76, 90; polarization, 264, 271–4; quantum computing, 278; quantum cryptography, 271–4; quantum teleportation, 274, 276, 277; Schrödinger's work, 102; Solvay Congress (1927), 146; term, 109
photosynthesis, 284
Physica, 199
Physical Review, 130, 177, 265
Physical Review Letters, 274
Physics, 265
Physics and Beyond (Heisenberg), 112–13, 137
Physics Institute, 15
Physics World, 281
pilot wave model, 152, 260–1
Pisa, 245
Planck, Max: career, 70, 148, 156, 256; discovery of "energy elements" (quanta), 73, 74–5, 83, 84, 90; honours, 78, 189; influence, 58, 63, 66, 109; Nobel Prize, 78; relationship with Schrödinger, 151–2, 158, 194; Solvay Congress, 146; successor at Berlin, 148, 151–2; work on black body radiation, 68, 70–4, 86, 87, 89, 108; work on electromagnetic radiation, 22, 67, 76
Planck's Constant, 73, 78, 119, 128, 140–2
Podolsky, Boris, 177, 265
Poincaré, Henri, 40
Pontifical Academy of Sciences, 189, 196–7, 258
positron, 159–60
Princeton: Bohm's dismissal, 260; Einstein Archive, 180; Einstein at, 5, 177; Feynman at, 162–3; Institute of Advanced Study, 203, 214; Schrödinger's lectures, 173–4
Principia (Newton), 27, 36
probabilities: Born's work, 135–6, 228; Copenhagen Interpretation, 144–5, 161, 162; de Broglie's work, 152; Einstein's work, 87–8, 122, 144; Heisenberg's work, 142; quantum world, 3, 122; Schrödinger's work, 88, 122, 135–6, 144, 163, 222, 228; statistical rules, 41, 164
probability wave, 144, 145, 153, 162
Proceedings of the American Philosophical Society, 181
Proceedings of the Cambridge Philosophical Society, 179
Proceedings of the Royal Irish Academy, 205
Proceedings of the Royal Society, 120, 134
proteins, 226–7, 235, 240–1
Prussian Academy, 160, 166–7

quanta: Bohr's work, 83–5, 86; Einstein's work, 67, 75, 76–8, 86, 127; light, see photons; Millikan's experiments, 77–8; Planck's energy elements, 67, 73–5
quantum chemistry, 207, 227, 228–30
quantum computers, 278–81, 287
quantum cryptography, 270–4
quantum entanglement, *see* entanglement
quantum jumps, 84–8, 114, 115, 135, 136–7, 140, 220
quantum mechanics: Aspect's experiments, 268–9; Bell's work, 262, 264–5, 266; Bohm's work, 260–1; Born and Jordan's work, 119, 132; Copenhagen Interpretation, 165, 177, 262; Cramer's work, 162, 163; development, 231; Dirac's work, 119–20, 134, 159–60; Eddington's work, 189; Einstein's work, 177–9; Heisenberg's work, 117, 118, 132–3, 138, 139, 140–2; Innsbruck meeting (1924), 103–4; interpretations of, 165, 177, 218, 219, 222, 228, 280; Many Worlds Interpretation, 222, 280; and reality, 288–9; Schrödinger's cat, 181–2; Schrödinger's work, 104, 109–11, 120, 128–30, 132–3, 150, 160–2, 163, 177, 181–2, 205, 218, 219–22, 234, 254–5; Solvay Congress (1927), 146–7; superposition of states, 221; term, 102, 103–4, 139; transactional interpretation, 163–5; transformation theory, 134
quantum numbers, 85, 87–8, 113, 120, 129
quantum physics: absurdity of, 5; accuracy of, 90; archives, 52; Bohr's work, 84, 112–13, 136–7, 138; "central mystery," 31; chemistry, 227; chess board analogy, 116–17; de Broglie's work, 127–8; development of, 2–3, 31, 35–6, 78, 108, 127, 155, 259, 269; Einstein's work, 3–4, 86–8; first version, 16; Heisenberg's work, 112–13, 138; lasers, 89; and reality, 281–4, 288–9; Schrödinger's work, 3, 5, 61, 102, 109–11, 136–7, 227; second quantum revolution, 66, 88, 91, 101
quantum reality, 146, 287–8
quantum revolution: first, 66, 67, 87, 146; second, 4, 30, 66, 88, 90, 91, 101, 108, 121, 146–7
quantum spin, *see* spin
quantum states, 116–17, 221–2, 232, 276–7, 289
quantum statistics, 108–11, 114, 136
quantum teleportation, 274–8
quantum theory: birth of, 22, 37; Bohr's work, 137; Clauser's experiments, 266; cosmology and, 189–90; education in, 95; Einstein's work, 88, 94; founding fathers, 82; Heisenberg's work, 113, 137; Rudolph's work, 287; Schrödinger's work, 58, 96, 101, 111, 137, 205; statistical approach and, 39, 142
Quantum Theory (Bohm), 260
Quantum Theory and Measurement (ed. Wheeler and Zurek), 181
quantum uncertainty, 140–2

radiation: alpha and beta, 79–80; background, 51; behaviour of, 39; black body, 16, 67–9, 70–2, 86, 87, 89, 108; Bose's work, 108; cavity, 68–9, 71; Doppler effect, 102; Einstein's work, 75–6, 77–8, 87–9; Feynman's work, 162–3; gamma, 79; Planck's work, 22, 67, 70–4, 76, 86–7; resistance, 162–3; Rutherford's work, 79–80;

radiation (*cont.*)
Thomson's work, 79; ultraviolet, 232; Wheeler–Feynman theory, 163
radio waves, 37
radioactivity, 78–9
Rathenau, Walther, 105
Ratnowsky, Simon, 95
Ray, Lucie, 215
Rayleigh, Lord, 71, 74
Rayleigh–Jeans Law, 71–2
reality: Bell's work, 264–5; Bohm's work, 288; Bohr's view, 145, 146; Copenhagen Interpretation, 178; Einstein's view, 139–40, 178–9, 288, 289; EPR Paradox, 177, 178; Heisenberg's view, 139–40; local, 179, 221, 264–5, 266; many worlds, 222, 280; of measurements, 115–16; quantum physics and, 281–4, 288–9; Rudolph's work, 287, 288–9; Schrödinger's view, 247, 249–50, 279, 280, 284; Vedantic vision, 61, 284
redshift, 199
Reichelt, Hans, 192
relativistic hydrogen equation, 125, 128, 130
relativity: general theory, 48, 58, 85–6, 94, 95, 189, 210–11, 215, 218, 248; Schrödinger's work, 58–9, 101, 210–11, 215, 217–18, 248; special theory, 3, 22, 44, 75, 93–4, 101, 102, 104, 125, 129; theory, 153, 159, 178, 179, 267
Rella, Lotte, 20
Rella, Tonio, 20
Reviews of Modern Physics, 78
Ribbentrop, Joachim von, 195
RNA, 226, 234
Rosen, Nathan, 177, 265
Royal Irish Academy (RIA), 205
Royal Society, 25, 120, 134, 216–17
Rudolph, Terry (Schrödinger's grandson), 285–90
Russell, Ann (née Forster, great-grandmother), 9, 18
Russell, Bertrand, 216
Russell, Emily (Minnie, grandmother), see Bauer
Russell, Linda Mary Therese (daughter), 213–14, 216, 286–7
Russell, William (grandfather), 9
Russell, William (great-grandfather), 9
Rutherford, Ernest, 79–82, 86, 189

Salpeter, Jakob, 46–7
Salten, Felix, 54
Santander, 175–6
"scattering" experiments, 89
Scherrer, Paul, 103
Schiele, Egon, 21
Schopenhauer, Arthur, 4, 60
Schrödinger, Anny (née Bertel, wife): in Arosa, 97, 126; depression, 216, 244; in Dublin, 204; early relationship with Schrödinger, 52, 54, 57; employment, 61–2, 170; engagement, 61–2; flight from Nazis, 196–8; friendships, 51, 131, 173; health, 216, 246, 253, 257; holidays in Ireland, 206, 210; holidays in Italy, 169, 245, 255; holidays in Tyrol, 136, 169, 245, 246, 248, 254; husband's death, 257; Irish citizenship, 215; London flat, 183; marriage, 62–3, 103, 132, 169–70, 183; mother in Vienna, 169, 189, 195; mother-in-law's illness and death, 64; in Oxford, 170–1, 175; postwar travels, 214–15; relationship with Hilde, 170; relationship with Ruth, 175, 189, 216; relationship with Weyl, 103, 126, 176; return to Vienna, 247–8, 251–2; social life in Zürich, 103; in Spain, 176; suicide attempt, 216; US trip, 149–50

Schrödinger, Erwin: family background, 7–10; birth, 10; childhood, 9, 10–13; education,

12, 18–20, 21, 22, 38, 43, 44–7; military training, 47–8; assistant to Exner, 48, 51; research, 48–9; *Privatdozent* appointment, 49; Congress of Vienna (1913), 52–3; first lecture course, 53; war service, 54–8; first papers (1914), 55–6; papers (1917), 58–9; engagement, 61–2; father's death, 62; Jena assistant professorship, 63, 64; marriage, 62–3, 103, 157; papers (1920), 63–4; Stuttgart associate professorship, 64; mother's death, 64; Breslau professorship, 65; Zürich professorship, 65–6, 91, 96, 98–9; TB and convalescence, 96–7; inaugural lecture at Zürich, 99–101, 102; papers (1922), 97–8, 101, 102; papers (1922–26), 101; Solvay Congress (1924), 101; life in Zürich, 103; Innsbruck meeting (1924), 103–4, 109; Mount Wilson experiment controversy, 104–6; *Meine Weltansicht (My View of the World)*, 106–8; work on quantum statistics, 109–10; wave equation, 111, 125, 128–9, 144, 159, 229; wave mechanics, 126, 129, 158, 228; Zürich colloquium (1925), 124–5; Zürich colloquium (1926), 129; papers on wave mechanics (1926), 129–30; relationship with Ithi, 131–2, 157–8; response to Heisenberg's matrix mechanics, 132–5; discussions with Bohr and Heisenberg, 136–8; response to Heisenberg's Uncertainty Principle, 142; US tour (1927), 149–50; Berlin professorship question, 147, 148–9, 150–2; Solvay Congress (1927), 146–7, 152–3; Berlin professorship, 155–7; Prussian Academy of Sciences, 158; papers (1930, 1931), 160; paper "On the Reversal of Natural Laws" (1931), 160–1; relationship with Hilde March, 168, 169–70, 174; decision to leave Germany, 168–9, 192; Oxford appointment, 170–1; Nobel Prize (1933), 121–2, 171, 172; Princeton lectures, 173–4; family life in Oxford, 174, 175; birth of daughter Ruth, 175; lectures in Spain, 175–6; resignation of Berlin professorship, 176; "cat in the box thought experiment," 179–82; Edinburgh professorship question, 184–5; Graz professorship, 185–6, 187, 190; family life in Graz, 188–9; interest in cosmology, 189–90, 199; letter recanting opposition to Nazism, 192–4, 198; dismissal from Viennese post, 194–5; dismissal from Graz, 195–6; flight from Nazis, 196–8; arrival in Oxford, 198; Dublin offer, 198, 204; Ghent visiting professorship, 198, 199; Ghent honorary degree, 200; arrival in Dublin, 200; Dublin Institute for Advanced Studies, 201, 204, 206–7; family life in Dublin, 204–5, 208, 210, 211–14; DIAS international colloquium, 209–10; search for unified field theory, 210, 214, 215, 218; birth of daughter Blathnaid Nicolette, 212; birth of daughter Linda Mary Therese, 213–14; postwar travels, 214–15; wife's suicide attempt, 216; health problems, 216, 218–19; Fellow of Royal Society, 216–17; at Innsbruck University, 217–18; work on interpretation of quantum mechanics, 219–22; *What Is Life?*, 214, 225, 226, 227, 234–40; return to Vienna, 243, 245–7;

Schrödinger, Erwin (*cont.*)
inaugural professorial lecture, 247; daughter Ruth's marriage, 247; life in Vienna, 248, 251–2; birth of grandson, 252; last lecture, 253; retirement, 254; ill health, 255–6, 257; death, 242, 257; grave, 257–8; scientific legacy, 259
FAMILY: children, 175, 212–13; father, 7–8, 19, 50, 61, 62; grandchildren, 247, 252, 285–7; mother, 7–8, 64; wife, 62–3, 103, 157
FINANCES: after First World War, 59; army pay, 59; attitude to financial security, 64–5, 173–4, 251; difficulties during period of inflation, 64–5; Edinburgh offer, 185; flight from Nazis, 196, 197; ICI funding, 170, 183; Irish income, 205, 206; Jena University salary, 62; money kept in Sweden, 172, 184, 194; Nobel Prize, 172; in old age, 251; Princeton lecture fees, 173; Vienna University income, 61, 62, 185, 246; Zürich University salary, 65–6
HEALTH: APPENDICITIS, 218–19; bronchitis, 96, 219, 246, 252, 254; cataracts, 216; declining health, 244, 248, 251; phlebitis, 246, 253–4; pneumonia, 253; sleeping pill overdose, 246; tuberculosis, 59, 96–7, 98, 101, 255–6
HONOURS: Berlin professor emeritus, 176; Fellow of Royal Society, 216–17; Ghent honorary degree, 200; Nobel Prize, 121, 171, 172, 196; Pontifical Academy of Sciences, 189, 196–7, 258; "Pour la Mérité" award, 253; Prussian Academy of Science, 158; TCD honorary doctorate, 206; torchlight procession, 151; Vienna University professor emeritus, 254
LECTURES AND TALKS: BBC talks (1949, 1950), 217; "The Crisis of the Atomic Concept," 247; "Elementary Wave Mechanics," 175; "Equality and Relativity of Freedom," 176–7; "Expanding Universes," 248; "The Fundamental Idea of Wave Mechanics," 172; "General Relativity," 248; "Interference Phenomena of X-rays," 53; last lecture, 253; "Nature and the Greeks," 215; on quantum mechanics, 205; Spanish tour, 175–6; "The Spirit of Science," 214; Tarner Lectures, 248; US tour, 149–50; on wave mechanics, 136, 149; "What Is Life?," 211, 224–5; "What Is a Physical Law?," 99–101
LOVE LIFE: adolescent crush on Lotte, 20; Anny, 61, 103; diary entries, 126, 169, 212; Ella Kolbe, 46; Felicie Krauss, 50–1; Hansi Bohm, 170, 183, 185, 193–4, 198, 214, 215, 216; Hilde March, 157, 168, 169–70, 174, 184, 189, 190–1, 195, 200, 204; importance to his scientific creativity, 126; Ithi Junger, 131–2, 157–8; "Kate Nolan," 213; Lucie Ray, 215; notebook record of lovers, 57, 61; Sheila May, 211–13; unknown girlfriend from Vienna, 126
PERSON: APPEARANCE, 5, 48, 52, 156–7, 185; interest in heredity, 224; languages, 11, 20, 210; philosophical studies, 60–1, 224; political views, 105–6, 190, 192–4; religion, 10, 18–19, 209, 224, 237–8, 250; reputation, 104–6; theatre-going, 9, 20–1, 208, 248

WRITINGS: "Are There Quantum Jumps?," 220; *The Interpretation of Quantum Mechanics* (ed. Bitbol), 219; letter to Synge on quantum mechanics, 254–5; *Meine Weltansicht* (*My View of the World*), 106–7, 256; *Mind and Matter*, 248–51; "On the Conduction of Electricity on the Surface of Insulators in Moist Air," 47; "On Determinism and Free Will," 237–8; "On the Reversal of Natural Laws" paper (1931), 160–1; papers (1914–15), 55–6; papers (1917), 58–9; papers (1920), 63–4; papers (1922), 97–8, 102; papers (1922–6), 101; papers (1926), 126, 129–30; papers (1930, 1931), 160; paper (1939), 199; poetry, 4, 212, 217; "The Present Situation in Quantum Mechanics," 181; *Space–Time Structure*, 215; *Statistical Thermodynamics*, 215; "An Undulatory Theory of the Mechanics of Atoms and Molecules," 130; "The Visual Sensations," 64, 101; *What Is Life?*, 4, 107, 214, 225, 226, 227, 234–40, 242, 249; "What Is Real?," 107–8

Schrödinger, Georgine (Georgie, née Bauer, mother): birth, 9; death, 64; family background, 8–10; funeral, 66; health, 59; holiday in England, 18; home in Vienna, 12, 64; marriage, 7, 10
Schrödinger, Josef (grandfather), 8
Schrödinger, Maria (grandmother), 7
Schrödinger, Marie (aunt), 7
Schrödinger, Rudolf (father): birth, 7–8; career, 8, 10, 50, 59; death, 62; family background, 7–8; finances, 59, 61; health, 61; influence on son, 45; life in Vienna, 12, 14; marriage, 7, 10
Schrödinger's Kittens (Gribbin), 162, 270, 273
Schrödinger's Philosophy of Quantum Mechanics (Bitbol), 220
Schulhof, Alfred, 208
Schuschnigg, Kurt, 188, 190, 191
Science, 233
Scientific American, 281, 283
Soddy, Frederick, 80
Solvay, Ernest, 146
Solvay Congress: (1924), 101; (1927), 146–7, 152–3, 156; (1933), 170; (1948), 216
Sommerfeld, Arnold, 104, 112, 113, 124–5
Space–Time–Matter (Weyl), 95
spectroscopy, 83
spin, 113–14, 125, 129, 159, 178, 264
"spooky action at a distance," *see* action at a distance
Stalin, Joseph, 208, 244
Stanford Linear Accelerator Center (SLAC), 263
statistical approach: Bell's work, 264; Boltzmann's work, 39–41, 44, 72, 99; Born's work, 144; Bose–Einstein statistics, 108–9, 114; Copenhagen Interpretation, 145–6; Einstein's work, 55, 87, 95; Fermi–Dirac statistics, 109, 114; Heisenberg's work, 142; Maxwell's work, 37; Planck's work, 67, 70–2; quantum statistics, 114; Schrödinger's work, 55, 99–102, 109, 135–6, 160, 215, 250, 253
statistical mechanics, 39–40, 44, 95, 101–2, 109, 160
Stefan, Josef, 16, 17, 21, 38, 41, 44
Stefan–Boltzmann Law, 16
Stuttgart, 64
"substitution," 133–4
superposition, 178, 179, 180, 221, 278, 282
Synge, John, 254

Tarner Lectures, 248

teleportation, 269, 274–8
thermodynamics: Boltzmann's work, 22, 39–41, 70, 72, 99; Loschmidt's work, 16; Maxwell's work, 70; Planck's work, 70, 71; Rudolph's work, 287; Schrödinger's work, 58, 160, 215, 236–7; second law, 40, 41, 70, 93, 99; statistical, 70, 160, 215; Stefan's work, 16
Thirring, Hans, 48, 63
Thomson, J. J., 79
time: absolute, 27, 29; "arrow of," 99, 287–8; Cramer's work, 162, 163–4; Einstein's work, 58, 88; hidden variables theory, 261; Newton's view, 27; reversibility, 40, 160–1, 162; Schrödinger's work, 100, 160–1, 250–1; statistical interpretation of, 250
Timofeev-Ressovsky, Nikolai, 230
transactional interpretation, 165
transformation theory, 134
Treatise on Electricity and Magnetism (Maxwell), 36
Trimmer, John, 181

UK Atomic Energy Research Establishment, 262
Ullmann, Elisabeth, 252
ultraviolet catastrophe, 69, 70, 71, 73
unified field theory, 214, 215, 218

Vedanta, 4, 61, 106–7, 256, 284
Vedral, Vlatko, 283–4
Vernam, Gilbert, 270
Vernam cipher, 270–1
Vienna: Allied occupation, 243; art, 21; blockade (1918–19), 60, 61; climate, 251; Congress of (1913), 52–3; history, 13–14; Hitler's entry, 190; Nazism, 190–1; physics, 15–17; quantum cryptography, 274; Schrödinger Archive, 149; Schrödinger's childhood, 11–13, 14–15; Schrödinger's education, 18–20; Schrödinger's retirement, 5; Schrödinger's return to, 218, 242, 245, 247–8; social life, 190–1, 248, 252; theatre, 20–1, 248
Vienna, University of: Boltzmann's work, 38, 39, 41, 42, 43; Doppler's work, 15–16; Loschmidt's work, 16, 17; Physics Institute, 15, 16; Rudolph's career, 287; Schrödinger's career, 48, 49, 57–8, 62, 184, 185, 194, 245–6, 247–8, 251, 253–4; Schrödinger's studies, 43, 44, 75; Stefan's work, 16, 17
Volta, Alessandro, 144
von Neumann, John, 261–3

Walton, Ernest, 209–10
Washington conference (1946), 238–9
Watson, James, 4, 239, 240–1
wave equation: classical mechanics, 125, 128; control of wave function, 221; Copenhagen Interpretation, 144; Maxwell's work, 128; "probability wave," 144; Schrödinger's work, 125, 128–9, 133, 136, 144, 152–3, 160–2, 261; spin issue, 159; time and, 160–4; use of, 229
wave function: Born's work, 135–6; collapse of the, 145, 153–4, 165, 177, 178, 179, 180, 181–2, 221–2, 261, 278, 280–1, 284; control of, 221; Copenhagen Interpretation, 144–5, 161, 281; de Broglie's interpretation, 152; entangled, 221; Schrödinger's work, 135–6, 161–2; superposition of, 180, 221; time and, 161–2
wave mechanics: Bohm's work, 260–1, 263; de Broglie's work, 158; matrix mechanics and, 133–5, 139, 221; quantum mechanics, 139; Schrödinger's

lectures, 136, 149, 150, 172, 175; Schrödinger's work, 118, 123, 126, 129–30, 132, 148, 151, 152–3, 158, 228; status of, 151, 154–5, 156, 227, 229
wavelengths, 69–70, 71, 73, 76, 102, 127–8
waves: "advanced" and "retarded," 163–4; amplitude, 135–6; de Broglie's work, 110, 124, 260; electromagnetic, 35, 37, 69, 72, 74, 128; Huygens' work, 30; imaginary numbers, 98; light, 30, 31–2, 35, 63, 71, 110, 128; Maxwell's work, 35, 63; particles and, 3, 77, 90, 110, 125, 128, 135–6, 143, 145, 228, 247, 255, 261; "phase space," 153; pilot wave model, 152, 260–1; Planck's work, 67, 72–4; probability, 144, 145, 153, 162; radio, 37; Schrödinger's work, 35–6, 63, 100, 102, 110–11, 247; Sommerfeld's work, 125; time and, 160–4; wave packet, 143; Wollaston's work, 82; Young's work, 31–2, 63, 77, 82, 102
Weill, Kurt, 155
Weinberg, Steve, 281
Weyl, Hella, 103
Weyl, Hermann (Peter), 95, 97, 100, 103, 126, 169
What Mad Pursuit (Crick), 239
Wheeler, John, 162, 163
Wheeler–Feynman theory of radiation resistance, 163
Wien, Wilhelm, 71, 74, 112, 135
Wien's Law, 71–2
Wilson, Charles, 138
Wisconsin, University of, 149
Wisdom, John, 248
Wittgenstein, Ludwig, 60
Wittmann-Denglass, Anton, 8
Wolfke, Mieczyslaw, 95
Wollaston, William, 82
Wooters, William, 269–70
Wren, Christopher, 26

X-ray crystallography, 49, 53, 94, 241
X-rays, 79, 89, 227, 230, 232, 240

Young, Thomas, 31–3, 63, 77, 82, 102

Zeilinger, Anton, 274, 277, 287
Zeitschrift für Physik, 108–9, 117, 141–2
Zimmer, Karl, 230
Zurek, Wojciech, 269–70
Zürich: ETH (Eidgenössiche Technische Hochschule), 75, 92–6, 124, 151, 154; history, 91–2; Schrödinger in, 64, 66, 78, 90, 96, 103–4, 108, 129, 150–1, 214; University of, 65–6, 92, 94, 103, 118, 151, 214

John Gribbin gained a PhD from the Institute of Astronomy in Cambridge (then under the leadership of Fred Hoyle) before working as a science journalist for *Nature* and later *New Scientist*. He is the author of a number of bestselling popular science books, including *In Search of Schrödinger's Cat*, *In Search of the Multiverse*, *Science: A History*, and *The Universe: A Biography*. He is a Visiting Fellow at the University of Sussex and in 2000 was elected a Fellow of the Royal Society of Literature.